Hubert Warren's cider press in Netherbury, West Dorset.

Published by Little Toller Books in 2022

Text © Liz Copas and Nick Poole

The right of Liz Copas and Nick Poole to be identified as the authors of this work has been asserted in accordance with Copyright, Design and Patents Act 1988

Jacket and endpaper photographs © Robin Ravilious and Common Ground

All illustrations © Liz Copas and Common Ground except: photographs by James Ravilious on pages 6, 26, 46, 118, 140, 158, 178 and 196 © Robin Ravilious and Common Ground; Dorset Sheet XXIX.SW on page 11 reproduced with the permission of the National Library of Scotland; Owl Jar on page 176 © Sherborne Museum; prize certificates on page 180 and Linden Lea; National Mark poster on page 187 courtesy of Bath Record Office Cider archives; Morse tokens on page 155 courtesy of Jim Chapman, Gloucester Orchard Trust; various postcards of Loders and archive information from the 1896 Tythe Apportionment of Loders, courtesy of Chuck Willmott of Loders History Group; various articles from Harry Warren's archive, courtesy of Alasdair Warren; haymaking at Ruscombe on page 17 courtesy of Graham Davies, Lyme Regis Museum; extract from *Cattistock – A Dorset Village 1916–2006*, courtesy of Buddy Langford.

Typesetting and design by Little Toller Books

Printed in Cornwall by TJ Books

All papers used by Little Toller Books are natural, recyclable products made from wood grown in sustainable, well-managed forests

A catalogue record for this book is available from the British Library

ISBN 978-1-915068-09-5

Traditional orchards and their apple varieties are part of the landscape heritage of the Dorset Area of Outstanding Natural Beauty (AONB); they are part of what makes this nationally important landscape special. Orchards are significant for their historic, cultural, wildlife and rare crop biodiversity. The Dorset AONB Partnership has supported the publication of this book to celebrate them and shine a light on their importance.

COMMON GROUND

The Lost Orchards

Rediscovering the forgotten
cider apples of Dorset

LIZ COPAS

with

NICK POOLE

LITTLE TOLLER BOOKS

We dedicate this book to the late Rupert Best, a great protagonist of all things Dorset, our long-time friend and supporter, and all round Good Egg. He was an enormous help and encouragement throughout the project, taking a keen interest in all our finds. Thanks to his generosity, we have had the luxury of watching our trees blossom into maturity on the land at Linden Lea orchard and established a mother-tree source for Dorset's future cidermakers.
It is sad that he is not here to see the project's finale.

In 1986, Common Ground started work on **Trees, Woods and the Green Man**, *a project which invited different artists, sculptors, photographers, illustrators, poets, cartoonists, playwrights and writers to explore the natural and cultural value of trees, and worked to deepen popular concern and practical caring for trees by publishing pamphlets, a newspaper called* PULP! *and several books, as well as commissioning plays and initiating art exhibitions at Tate Britain, the Natural History Museum and The South Bank Centre. During 1989 and 1990, Common Ground commissioned James Ravilious to photograph the orchards of the West Country. These photographs helped Common Ground launch various campaigns, including* Save Our Orchards, Apple Day *and the* Community Orchard *project. The colour photographs featured in this book have been selected from this archive.*

Contents

When leaves that leätely wer a-springen
Now do feäde 'ithin the copse,
An' painted birds do hush their zingen
Up upon the timber's tops;
An' brown-leav'd fruit's a-turnen red,
In cloudless zunsheen, over head,
Wi' fruit vor me, the apple tree
Do leän down low in Linden Lea.

William Barnes, 1859

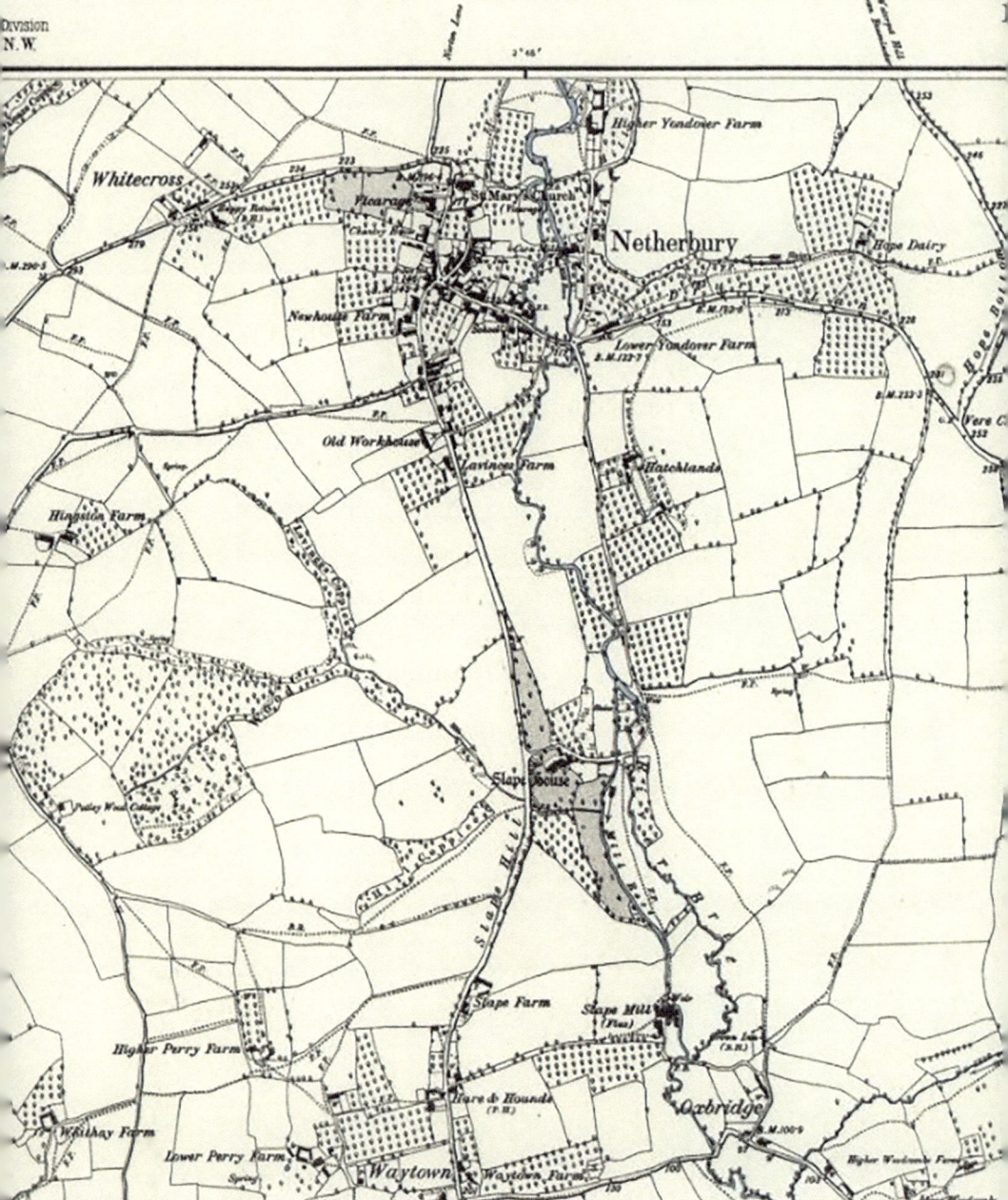

Dorset Sheet XXIX.SW, surveyed in 1886 and published in 1888, shows the abundance of small orchards before the start of the twentieth century, where much social change shifted the use of the land.

Celebrating Small Orchards

Common Ground

About two-thirds of Britain's orchards have been lost since 1960. Dorset once boasted over ten thousand acres of orchard, with 102 acres of trees in one, small West Dorset parish (Chideock). There were no major commercial cider producers in Dorset, like those that had established in Somerset, Devon and Herefordshire. There were plenty of breweries (beer was the favoured drink in many places, especially to the eastern side of the county). But old maps show that every small Dorset village was surrounded by orchards. Although records of the apple varieties that were used to make local cider are scarce, each farm and cottage had its own trees and the farmhouse cider of each village would have had its own local character that suited the taste of its residents.

Two World Wars brought huge social changes, and with them many orchards disappeared under the plough as land was cleared for growing crops to feed and house the growing nation. The only memory of the trees that once thrived are in the names of roads and houses: Orchard Close, Old Orchard Way, Apple Tree Cottage. The industrialisation of agriculture changed local farming practices forever, leaving less time for harvesting and pressing apples, while Government subsidies and grants over the last 60 years put an end to the traditional orchards as farmers were paid to clear trees and grub-up hedgerows. Of the many thousands of dessert, cooking and cider varieties which were once grown across the UK, only a few handfuls are widely known and used today.

The apple is a wonderful symbol of variety and meaning; the orchard a rich example of our cultural landscape. Together, the apple and the orchard provide a way of expressing both the robustness and the vulnerability of our local landscape and culture. With the loss of an orchard goes the loss of ecological diversity and the loss of knowledge of recipes, songs, customs, wassailing, cidermaking. Lost too are the social gatherings for work and pleasure, the sharing of knowledge about the place and the sharing of skills of pruning, grafting and growing. The intricacy of a community and its distinctiveness is diminished if orchards and apple varieties are lost.

Orchards are still being grubbed up and replanted with cereals or ousted by new developments for cars, trains or housing, while many orchards simply fade with neglect. Thankfully, the work of Liz Copas and Nick Poole is now supporting the revival of interest in small-scale orchading and cidermaking. Their searches since 2007 have unearthed many interesting and unique apple varieties that are now ready to be reintroduced into orchards. In rediscovering the last vestiges of Dorset's lost orchards, they have begun a hopeful story that will inspire others to look closely at their surroundings and take steps to celebrate and conserve the trees and orchards that make their locality special.

COMMON GROUND
commonground.org.uk

Apple guardian by Clifford Harper.

A cider jar from Warren's of Netherbury, West Dorset.

INTRODUCTION

In the Cider Shed
Nick Poole

Cidermaking is not just all about the technical art and skills of a Master Craftsman producing the finest quality drink he or she can. There is an even more interesting and complex history in the story of cider that doesn't come anywhere near a commercial product, and is often hidden from general view behind the creaky old doors of a cider shed or at the back of the farmer's barn. This is where the real world of cider-lore, where generations of country folk have gone for solace and comfort and taken from that dusty barrel in a dark corner a small sup of relaxation and pleasure.

This escapism was once a reward and comfort to farm labourers and country workers who toiled on agricultural land, and who even received cider as part of a weekly wage. Beginning the day at 6 a.m., with an early morning sun brightening the gloom of the dusty cider shed, they would fill a four-pint firkin and strap it to their belts as they made their way to the fields. It was what they relied on for refreshment and energy, keeping their spirits raised, as they became part of the 'crew' with a shared sense of endeavour, of challenge and achievement as the day's work would grind on. These days, we imagine them all drunk by lunchtime. But the hard work sweated out the alcohol and kept them going through a twelve-hour day – the cider ration was doubled to eight pints during haymaking season and harvest time.

My father, who served a farm apprenticeship in Devon before the Second World War, had the daily duty (at sixteen years old) to dole out the cider to the old hands, and said that most of them would down a pint before they left the shed with their full firkins. By this time, including cider as part of the wages had long since vanished. Yet the workers were still given free cider rations on many farms, and the better the cider the happier the work force.

At the end of the Second World War, Britain changed immensely and working practices that had altered little for generations suddnly disappeared, with new machinery and modern engineering taking the place of horses and hard grunting. The numbers of farm labourers diminished hugely as mechanisation swept across the agricultural world. Where idle times at the end of harvest were once filled with the sound of the workforce turning their hands to cidermaking, the orchards and cider barns gradually fell silent.

Hand-turned *scratter* mills were abandoned to rust and rot, along with many other outdated machines, lost in obscure corners of the farmyard gradually consumed by nettles and ivy. The presses, too, became idle and forgotten in gloomy sheds, or else sold as garden features to gradually rust away in the rain.

Life was changing rapidly in the countryside. As farm workers lost their jobs to machines, they moved from the villages and the small family farms increasingly sold up, no longer able to compete with the large-scale farmers who intensified their production methods and increased their profits. These new farmers no longer considered the annual apple harvest and cidermaking a worthwhile reward, and consequently turned old orchards into arable lands or pasture. Cidermaking was pushed to the margins, supported only by the old die-hard drinkers who had grown up on it and couldn't do without it. In West Milton, where I grew up and still live, the last cider shed to satisfy the village thirst, at Lynch Farm, had fizzled out by the end of the 1950s.

Besides once being an integral part of working life, the cider shed was also a place of relaxation and escape from the day's toil, somewhere to enjoy a drop of alcohol that was free and easily accessible. A place to socialise and unwind, gossip, sing and swap jokes, compare the tribulations of the day, to sympathise and

Haymaking at Ruscombe, West Milton, in the 1930s. Note the jars of cider in hand.

commiserate, encourage or disagree. For many on low wages cider sheds provided respite. They were a place to regenerate, be comforted by a shared destiny, and feel part of the great rural working community.

Of course, this could only continue with the support of the patron, the farmer or the Lord of the Manor who provided the facility and apple crop to sustain the enterprise. So when they saw that it was no longer financially viable, or costing them money, the opportunities were withdrawn, and the era of the traditional cider shed slowly came to a close.

In parallel with this, times were also changing for the fortunes of commercial cider. By the 1970s a fashionable, new drink was hitting the English palate in the form of lager – the scourge of both the cider industry and the real ale brewers. Heads were being turned, young people were being enticed by this chilled, fluffy, golden beer that was too easy to drink, its inherent lack of flavour ignored because of its ability to quench a thirst on a hot summer's day. Indeed, its bland flavour was a great advantage in providing a new generation with a beer that wasn't challenging or debatable in the way that traditional ales and ciders had always been.

Some of the brewers fought back with kegged, pasteurised and fizzed-up ales, and so as far as flavour went, this became a race to the bottom – and the brewers still couldn't meet the challenge of lager. In the mid-1970s, cider also suffered a further setback when its duty was raised to the level of beer and what had once been a popular, cheap drink (approximately half the price of beer) went into a long period of decline.

Larger cider farms and industrial-scale cider producers rose to this challenge by offering lighter ciders, fizzy ciders, cider in cans and pressurised kegs of cider for the pub trade which helped them retain a margin of the drinks industry. But it wasn't until after the new Millennium, in around 2005, that the cider world began to reassert itself and make a real comeback. Magners, a little known cidermaker from Ireland, hit the British market running, with their 'cider on ice' marketing campaign, and within a year they were storming up the drinks charts.

As a drinker of traditional farmhouse-style ciders, I have been puzzled how Magners became so popular, even putting a dent in lager sales. But drinks are subject to fashion, and it looked like cider was coming back in grand style. Single-handedly, Magners rejuvenated the British cider market and others were quick to follow on, with production throughout England increasing every year. Even the major lager producers who became aware of their falling sales came out with their own brands of cider to compensate.

Cider was finally turning the tables on the drink that had once almost brought it to its knees. Bubbling along underneath all the high commerce of mass-produced ciders, a new customer was developing who was very interested in provenance and traceability. They wanted to know where their food and drink was coming from, what it was made from, how it was handled and how close it was to the origins of the product. For cider, this meant a resurgence of interest in the basic product, the traditional farmhouse cider, the unadulterated full-juice, full-flavoured, full-strength, blow-your-socks-off cider

What's behind the old oak door? Off Bluntshay Lane, near Shave Cross, here is perhaps the most famous cider shed in West Dorset: 'Will's Surgery' featured in the hilarious film made at the Dare family's farm in the 1960s by Clive Gunnell for Westward Television (now on YouTube). The 10 minute film shows the cidermaking process in the barn accompanied with tall tales and a raucous rendition of 'To be a Farmer's Boy'.

In the shed with Nettlecombe Cider Club.

that our rural forebears took for granted.

Of course, rural villages are no longer filled with farm labourers and the working classes of country life. These days villages are a diverse cross-section of people from all walks of life, with different backgrounds and interests. They may never have considered drinking cider, preferring wines and spirits, and may not appreciate a two-handled clay jar passed from lip to lip in the traditional way. But in the cider shed, cider is king, and it is now drunk from respectable glasses with an impressive array of foods shared out on the communal table. Everyone drinks the cider and murmurs of satisfaction and approval can be heard, and a pleasure is felt in the knowledge that everyone has had a part in this small enterprise.

Over the past twenty years, I have seen many groups form to make community cider and plant community orchards. The cider club has replaced the farmer's barn as a place to enjoy cider and it has become an important part of life for those involved. Sometimes it has been simply a few friends making cider in a garage or garden shed; sometimes bigger enterprises have emerged such as those at Chideock or around Monkton Wyld, where locals gather to press their juice and then take it home to ferment in their own ways.

The national revival of cider drinking has been a huge success over the past twenty years and in Dorset we now have more than 20 commercial producers. For me, however, the real enjoyment of cider comes from supping a glass in its natural environment, that quintessential place that links us to all the generations before: the cider shed. Of course, fashions change and other temptations will come and go, but I do feel that as long as there are people who seek the company of others and enjoy a shared sense of community and belonging, there will always be a cider shed and I hope that at some point everyone has an opportunity to experience this special drink in its unique setting.

A thirst for proper cider

I have drunk cider – 'proper cider'– for most of my life, starting as any farmer's son would do with the annual hay making, when barrels of cider were brought in to keep us going through those long, hot dusty days. I say *proper cider* because this was at a time in the late 1960s and 1970s, before the modernisation of this traditional drink had really made much progress; when you expected your cider to be hard, flat, slightly fruity and with a hint of razor blade at the back of the throat. This to us was an energy drink long before Red Bull or Monster appeared, and kept us working until long after dark, hauling in bales from the fields and stacking them tightly into cramped barns, with itchy straws and hay seeds mingling with the sweat and dust on your brow. The cider gently anaesthetised us to these discomforts as we toiled on till late, and we were always grateful for the bread and cheese plate that finish the evening – and then that final relaxing sup. We never felt inebriated in any way, mostly I think because we sweated it out. And that is how I suspect the farmworkers of old, with their four to eight pints a day, could also complete their chores without falling down drunk.

Our cider in those days was purchased from the Warrens of Netherbury, who ran our local commercial cider farm, and were very well known in West Dorset for producing a premium drink much appreciated by young and old alike. I came across Hubert Warren again in 2000. Just as I was beginning to make cider, he was in the process of shutting up shop as his sons had no interest in continuing the business, thus bringing to an end almost 80 years of

Nick contemplating how Matravers cider will taste.

the cider tradition. I was disappointed to see one of my childhood heroes disappearing. I was also very aware how few people were left in Dorset with the cider knowledge – there were no longer any serious commercial producers for the entire county. I could find no other commercial makers in Dorset, only a few old-time enthusiasts and the long-standing Chideock Cider Circle who had just turned 50 years old and were probably the biggest amateur producers in Dorset.

Inspired by their enthusiasm, I rented ground with an old cider orchard and asked my neighbours around West Milton if they would like to form a cider club. It would be the first time West Milton had had an active cider shed since the late-1950s, but the response was positive, and twenty years on it is still going strong as a very important part of our community.

A cider shed in a small rural village that has no shop, pub or

The last Golden Ball at Netherbury.

church, provides an important focus for the community. As it was over a hundred years ago, it is a place to meet up with friends and neighbours, to catch up on the gossip, and see other folk from the village that you would rarely meet if it wasn't for the shed. At harvest time, there is also a well-supported turnout to bring in the apples and make the cider. It gave our club members that shared sense of endeavour, a satisfying feeling of enterprise and achievement.

With no Google to help research, the knowledge of Hubert Warren and the members of Chideock Cider Circle was invaluable. We also gleaned what information we could from library books, and by the autumn of 2000 we made our first batch of cider. It was mainly from Morgan Sweet apples and other varieties that were unknown to me. We were delighted with the results and very excited that we had managed to produce something that was so pleasant to drink.

Encouraged by our early success, we continued to grow the West Milton Cider Club, and our knowledge of cidermaking improved with each year. I became so fascinated by the whole process that I found myself wanting to dig deeper and deeper into this curious business of turning simple apple juice into a first-rate alcoholic drink. Each time I dug, I started to understand the science – the fermentation, the storage, the hygiene – and began to realise that the fundamental success was always underpinned by that most important ingredient: the apples. The more books I read and the more cider I made, the more I began to appreciate the importance and significance of this.

Dr R. K. French's excellent book *The History and Virtues of Cyder* was a great inspiration that helped me look way beyond just producing a decent drink, and I realised I needed to get to know the apples of my own area in a much deeper way. In those early years, I had no idea what varieties had gone into the Club's cider, only that it was pleasant and drinkable. Our main orchard was at Maiden Crate in West Milton, where my wife and I were renting. But this only supplied us with apples for our first two seasons. I had been told that they were all cider apple trees, but not even our landlady knew what they were. Sadly, early in 2002, seven of the best cropping trees blew over in a severe gale, leaving us with just three very old full standards in production. And so began our annual scavenge for apples that gradually took us further afield.

I always insisted on collecting from orchards that had guaranteed cider apples, and not just cookers or eaters. But it soon became clear that the owners of orchards invariably had only one or two cider trees, surrounded mostly by new culinary and dessert varieties and they never knew what the cider apple trees were. It was at this point that I started packing up apple samples and sending them to the former Long Ashton Research Station in Bristol. It was here that I first met Liz Copas, just prior to the government's decision to close down the entire operation, after 100 years of very valuable and important research.

By now my thoughts had turned not just to the simple task of trying to find the best apples for our own cider. I wanted to know more about what type of cider might have been drunk in Dorset by previous generations. We were fortunate in West Milton that most of the old orchards contained well-known Somerset and Devon varieties, and with Liz's help we had soon identified Dabinetts, Chisel Jerseys, Yarlington Mill, Sweet Alfords, Browns and an abundance of Bulmer's Norman trees, many of which were planted in the hedgerows along Ruscombe Lane at the back of our village. It was obvious that at some point someone who understood good cider had invested wisely in their choice of apples, but sadly the history of who planted them – and why – had been lost with successive changes of ownership.

At this point, I naively assumed that most rural Dorset villages would be similarly blessed with examples of good quality cider trees, if you just knew where to look. I later came to discover that in fact West Milton was one of the best supplied villages for cider apples in Dorset!

Coinciding with my growing interest in this subject and by a stroke of good fortune (for me), Liz retired and moved to a place near Crewkerne, just over the Dorset-Somerset border. Hoping that she might now find herself at a loose end, I asked her if she would be interested in helping me carry out a bit of research into the lost cider orchards of Dorset, and see if we could establish what sort of cider people would have drunk 100 years ago. With her interest in history and passion for all things appley, she enthusiastically agreed. At least, that was how I interpreted it!

For several years, we gradually put together a picture of Dorset's cider apples, meeting some very interesting enthusiasts along the way and even discovering parts of the county that I had never visited before, despite having lived here my entire life. We attended many Apple Days and food festivals with our display boards and were surprised at how interested people were. Before long, we were travelling all over Dorset, following leads to overgrown orchards, corners of gardens, or to single trees lost in ancient hedgerows.

Searching, Grafting,
Planting, Tasting

Apple   Bellerwet / late. [illegible] conical [illegible] clubby green [illegible] flush

Best Bearer

Buttery d'Or   Sharp early. Pastry apple. medium [illegible] round. yellow. some russet veining

Golden Ball   (Nursethik, Polly. "Go Boys") (Yeovil to [illegible] Sharp. midseason. medium size, round yellow — some red flush.

Lucombe   Sharp. late. medium. round green. very hard texture. ("Morn Pippin"?)

[struck out]

Crimson King (Kings Favourite) rather conical crimson with some russet ~~marking~~ markings. midseason.

Long Stem   Sharp late. large conical red.

Round Apple   soft midseason. large flat. striped.

Somerset Bearbtin (Runaway, ~~[illegible]~~) soft. ~~medium~~ rather late medium size round to flat. striped & flush dark red.

Somerset Crab (Goat) Sharp. late. large conical green with some red stripe

Tom Legg   soft midseason medium — [illegible] yellow.

Mr Pickford's List

Liz Copas

The Long Ashton Research Station in Bristol started life in 1903 as the National Fruit and Cider Institute, and through its essential work, became the saviour of the then failing cider industry. During its time, it helped promote England to its place as the world's foremost cider producer. My job at Long Ashton with fellow scientist and cider pomologist, Ray Williams, was to find new and better ways of growing cider apples to the perfection expected by the modern industry towards the end of the last century. Much of the experimental work we did was about getting the trees to crop regularly, since our traditional cider varieties naturally and inconveniently tend to take every other year off. In later years, my work was more concerned with persuading orchard owners to practice some integrated pest management by reducing their spray chemicals and encouraging beneficial insects to do some of the crop protection work for them.

Throughout the century of its existence, Long Ashton gained an impressive reputation for making excellent cider with fruit harvested from the collection of traditional cider trees in the experimental orchards. The weekly bottled cider supplies invariably sold out rapidly, with regular deliveries made countrywide, even to the House of Commons. Scientists used to come to the Station from all over the country, ostensibly on scientific business, but never failed to visit the cider house to fill their car boots with demijohns or bottles of the best, clear, honey-coloured, still cider. Mine was a pleasurable, mostly outdoor job that took me to all parts of the cider growing counties, from Herefordshire down

Opposite: The notes of Mr Pickford, the Cider Instructor. Written in the 1930s, these writings were the only clues we had.

National Fruit and Cider Institute, 1904.

through Somerset, Dorset and Devon to work in orchards in all kinds of weather. Along the way, I learned to recognise the different faces and characters of cider apples and how their trees grew and responded to bush orchard cultivation.

During my time at Long Ashton, I came to expect an autumn visit from Nick Poole, who travelled from West Milton with a handful of assorted apples for me to identify. At the time, there seemed to be little general interest in finding a use for fallen apples, and many lay wasting in gardens at the end of each summer, but it was a good gleaning for a craft cidermaker like Nick. It was often a hard job to put a name to any of the apples he brought, since most were rather ordinary cookers or eaters. Occasionally, he would find a really promising fruit with a little flavour and perhaps a hint of astringency that might just make a good drop of cider. Back then, Nick was just starting on the road to becoming a professional cidermaker and was always experimenting with his apple finds, blending and mixing flavours.

There were few other cidermakers in Dorset at the turn of the last century, and Nick was alone in wanting to know what the real taste of traditional Dorset cider might have been in the past. He knew there had to be some proper cider apple trees out there, with good tasting apples that would make the right flavoured authentic blend. With so many small orchards lost or destroyed since the Second World War, it was clearly a time to foster interest in their worth. In 1796, Dorset was recorded as having 10,000 acres of cider orchards. Now, in the twenty-first century, travels around the county show few signs of orchard life compared with the abundance displayed in every village on the Ordnance Survey maps of 100 years ago. If there were lost trees and orchards out there, they needed uncovering and rescuing.

By the time Long Ashton closed its doors in 2003, I had unearthed a modest amount of historical writing about cidermaking in Dorset from the archives. But there was a big gap in historical records, and almost nothing to shed light on the decline of the county's farm cidermaking and the disappearance of all those thousands of trees. Through the 1920 and 1930s, however, Long Ashton initiated a farm Cidermaking Instruction Scheme for all the cider producing counties. The instructor, a Mr Phillips Pickford, travelled from farm to farm advising on orcharding, the best cidermaking methods and the most appropriate apples to use. His writing – a scanty few sheets – was the only clue I had to Dorset's own handful of cider apples.

With Mr Pickford's notes, Nick and I first set off around the Dorset countryside in the autumn of 2007 and saw quite a remarkable assortment of trees. Although some of these were in large, well-tended orchards that were still in full production, wherever we went there were no records or local knowledge about the varieties had been lost. Some were 'garden' trees, where newly-built housing had encroached onto old orchard land. Sometimes, to get to other trees and orchards, we fought our way through bushes and hedges. Now and again something very interesting would turn up. It was always much more exciting when we found a cider apple that couldn't immediately be identified, and even

A big old Reinette Obry tree at Muddicombe, the Best family's orchard in Melplash.

more promising if, after trawling the Long Ashton archives, we still couldn't put a name to it. We knew it would take some hard detective work before we could claim these apples as 'new' Dorset cider apples. But if they were not in any of the national archives, that is what they would become.

In 2008, we took what we had found – some 26 apple samples – to the lab at Thatchers Cider in Somerset for juice analysis, and also made several individual ciders for tasting ourselves. Most of what we made was fairly thin, lacking body and depth of flavour, but one or two samples proved to have some useful *tannin* and character. We felt it was unfair to judge the fruit from old, mostly uncared-for trees too harshly. The final judging could only come from cider made from clean, fresh fruit harvested from young, healthy bush trees.

Alongside the discovery of these apples, what we also found out was that Dorset's older orchards were neglected and in dire need of renovation. What we needed was to plant an experimental orchard of Dorset apples ourselves, taking what we'd discovered in the fields and hedges to a place where we could monitor and nurture a new generation of trees. This would be an orchard that

Saving some shoots from a Kings Favourite tree in Pymore.

A young Buttery Door in the Best's Muddicombe orchard, Melplash.

salvaged a lost heritage, rediscovered the local distinctiveness of apples, and put Dorset back on the map as a cidermaking county. All this, Nick and I hoped, would help encourage people to take better care of the older trees and orchards, while being useful to the new generation of small-scale cidermakers in Dorset.

Linden Lea: a Dorset trial orchard

Four years after Nick and I began our project, on 19th March, 2011, Henry Best (aged 3) helped plant a Golden Ball cider apple tree that had been propagated from a tree in Hubert Warren's orchard in Netherbury. This was the final tree planted at Linden Lea, a trial orchard at the centre of Melplash village in West Dorset. There was no other orchard like it, dedicated to the Dorset apple varieties that we had collected and rediscovered since 2007: Symes Seedling, Sweet Alford, Hunters Ground, Lancombe, Marnhull Mill, Dorset Winter Stubbard, Yaffle, Cattistock Pink, Golly Knapp, Ironsides, Tom Putt, Stubbard, Matravers, Puddletown, Best Bearer, Frome River, Buttery Door, Golden Ball, Marlpits, Dewbit, Tangy, Meadow Cottage, Tom Legg, Reinette Obry, Kings Favourite, Fillbarrel, Cap of Liberty, Hains Sweet and Hains No.2.

Alongside the making of Linden Lea Orchard, we were also able to distribute over 300 Dorset cider apple trees, all excellent young bush trees *grafted* and grown by John Worle in his nursery up in Herefordshire. These tree went to 30 new homes all over Dorset, many private orchards and some to the National Trust. There are now three collections: the complete complement in Melplash, another set at Thatchers Exhibition Orchard in North Somerset, and a third set at Liberty Farm, Closworth, Dorset.

Since 2011 our trees, planted in pairs at Linden Lea, have been allowed to grow in the fine Bridport soil. They have been left, more or less, to their own devices without pruning: we want to see how they grow naturally, and when they might begin to bear fruit. Given over ten years to mature and settle into their growth, the individual character of each *variety* has developed well. It

Cider to celebrate planting the final tree in Linden
Lea orchard in 2011. John Worle (with spade) and Liz
Copas (with cider).

is hard to imagine a more varied collection of apple trees. Few
have developed an ideal habit on their own, but this is typical of
most traditional cider varieties. Most would have benefitted from
some early tree training, to achieve an optimum shape and size.
A couple are poor, ungainly things, best left for a sympathetic
enthusiast to grow for their exceptional *vintage* fruit qualities.

Some of the varieties we planted had not been forgotten,
such as Kings Favourite and Golden Ball, both well known in
West Dorset. But some of the apples in Linden Lea could not be
identified and had to be given new names, like Marlpits (after the
name of the farm where it was discovered) and Dewbit (a Dorset
word for breakfast). A few years ago, our top fruit scientists at
East Malling Research Station in Kent initiated a public scheme
for subscribers to submit apple leaf samples for analysis, for a fee.

Linden Lea in full bloom.

Following a DNA search in the database at East Malling and the National Fruit Collection at Brogdale, each unique variety can now be registered as a Local Variety. Our Dorset finds are now certified, and their names registered.

Then a few years ago, Bristol University also initiated an independent genomics scheme. The university team sampled our Dorset finds and went on to introduce our apple trees into the National Fruit Collection in Brogdale. Between them, Bristol University and East Malling have developed some specialist software that can unravel the DNA genomes and reveal much about each variety's parentage and ancestry.

This has allowed Nick and I to search for clues in the genetic makeup of our Dorset varieties with some exciting results. All the young trees have had their DNA sequenced and all have proved to be unique when compared with known apple varieties in the National collections. Also, some of our Dorset finds have ancient lineage that links them to the earliest cider apples that were ever grown in this country, perhaps related to those that the French monks brought over at the time of the Norman Conquest – or even before that. This is exciting news for Dorset's cidermakers who can now boast that their cider not only tastes good and authentic but also has a history.

In their early years at Linden Lea some of the young trees gave a sprinkling of apples as they established themselves. More fruit came with each autumn that passed, but in 2019 plenty of them produced a really good crop. There was every colour and shape – from small, green, *sharp* apples like crabs, to large juicy red apples, good to eat from the tree. It was an excellent opportunity to taste a bite from each. That year, there was plenty for Nick and I to harvest from each tree as they began to mature throughout September and early October. Although Nick had to adapt his cider press for small individual batches of fruit, the juice flowed: some a clear honey colour, some dark pinkish brown and *sweet*.

A sample of each apple pressed went to Thatchers Cider laboratory, and they were kind enough to analyse the juice and confirm that the trees were developing some maturity. Over winter, the demijohns bubbled away with 22 apple varieties in Nick's cider shed at West Milton. By autumn 2019, they were being tasted.

The Taste Test

It is a well-known fact amongst cidermakers that most cider apples are best blended together, to balance out the inconsistencies that appear in the individual varieties. With this in mind, we enlisted prize-winning cidermaking guru Bob Chaplin (with his 46 years' experience of making cider for Gaymers and Magners) to help us with an initial tasting of the cider in a raw and immature state. Most were pleasantly balanced. Some were rather lacking in body. The hardest to appreciate were very sharp, full of *malic acid*, but perhaps useful for blending. Happily, the best were pleasant, moderately astringent and slightly spicy or fruity. The Linden Lea trees had provided us with a very varied and interesting collection, with unique flavours and aromas.

We set ourselves a hard challenge to decide which were good enough to go forward as single varieties in a *glass-to-glass* taste-off. Several of the sharper varieties would certainly have been welcome amongst barrels of *bittersweets*, but would not have been popular in a pint glass. It was very satisfying, however, to arrive at a figure of fourteen which we agreed would, with a bit of sweetening and time to mature, develop into some interesting single variety ciders. But what we needed was a real test.

On Saturday 12th September, 2020, a select band of Dorset's young craft cidermakers were hosted by Rupert Best in Hincknowle Barn, Melplash, to help taste and judge a set of unique single variety ciders. The now bottled and matured ciders were taken to the glass and thoughts and comments invited. All agreed that, as single varieties, they held their own, and if they were to be blended, could stand with the best of traditional cider apples from other parts of the West Country. A can be seen from the scores (*see* page 40), with the exception of Lancombe, there were no real screechers. The remainders were all quaffable and our observations led us to conclude that with mostly low tannin and minimum astringency these ciders would have wide appeal. At the end, on a final blindfolded tasting, it was unanimously agreed that the Golden Ball was the outright winner, justifying its history as one of Dorset's most historic and revered apples.

Serendipity or selection?

Cider apples come in many shapes and colours, but it is their juice and their differing flavours that puts them apart from the sorts of apples we like to eat. The biggest family of traditional West Country

All ready for the glass-to-glass cider tasting at Hincknowle Barn, 2020.

cider apples is the bittersweet. These apples have plenty of tannin, sugar for the fermentation and just a little acidity to help balance the juice. Tannins are complex *phenols* that evolved in wild apples as part of their natural plant defence mechanisms. To qualify as true a bittersweet, the juice should have more than 2 g/l of tannin and less than 0.45% malic acid, together with some sugar to ferment it. After the fruit is crushed ready for pressing, it is the oxidation of tannins that give the juice that familiar golden colour. The individual blend of phenolic compounds contributes to the traditional flavour, the 'body' and 'mouth-feel' of real cider.

Many bittersweets are offspring of the trees from neighbouring orchards, either selected deliberately or by chance from seedlings grown from pips in the *pomace* at the end of cidermaking. John Mortimer, a pioneering agriculturalist and Fellow of the Royal Society, describes in *The Whole Art of Husbandry* (1708) how after having made any 'Cyder', '*Verjuice*' or 'Perry' the pips were sifted out of the pomace with a riddle, then sown as soon as possible in the autumn,

Results of the Linden Lea taste test

SYMES SEEDLING

The initial taste was of quite high acidity but made more palatable by a good level of tannin. It had a light fruity background and texture that was approaching a wine-like quality.

DRINKABILITY: 8/10

MARLPITS

Smooth and mild mannered, being low in both tannin and acid. There was a hint of apricot and the flavour lasted well on the tongue. Very smooth and non-challenging to drink.

DRINKABILITY: 7/10

YAFFLE

The next-door neighbour of Marlpits and with a very similar character. There was perhaps a bit more mustiness and woody background to this one, with a slight hint of liquorice. Overall, an easy drinker.

DRINKABILITY: 7/10

LANCOMBE

The third in the trio from Marlpits Farm orchard was so high in acid as to be a challenging drink. It was perhaps promoted beyond its station as a single variety, but blended with the previous two it could add character and would sit comfortably in the orchard alongside the other two.

DRINKABILITY: 2/10

CAP OF LIBERTY

Quite sharp and astringent with a thin rather characterless background. Probably better blended with others. This is in fact a traditional Somerset apple that seems to have achieved some well deserved popularity in Dorset.

DRINKABILITY: 6/10

FROME RIVER

Unusually dark with hints of pink, suggesting high tannin but overall a good balance and pleasantly easy to drink.

DRINKABILITY: 9/10

HAINES SWEET

Almost wine-like quality, low in acid and tannin with a buttery feel. Easy to drink.

DRINKABILITY: 8/10

DORSET WINTER STUBBARD

This was the surprise of the evening, displaying a well-rounded smooth finish that was very easy to drink. A good level of soft tannin and a nice texture on the palette.

DRINKABILITY: 8/10

GOLLY KNAPP

This presented a bit more of a challenge, with quite high astringency and long dryness on the tongue, probably arising from a hard tannin background.

DRINKABILITY: 5/10

CATTISTOCK PINK

Very much a traditional farmhouse background that conjured up images of gloomy cowsheds. Quite complex and slightly challenging. Probably more enjoyed by old-time traditionalists.

DRINKABILITY: 5/10

GOLDEN BALL

A very well balanced cider, pale straw in colour with notes of honey and a light floral background. Sufficient sharpness to make it interesting but not overpowering.

DRINKABILITY: 9/10

MATRAVERS

Very light, tending towards thin, but with a nice fruity background with low astringency and quite easy to drink.

DRINKABILITY: 7/10

TOM LEGG

Similar to Matravers, with a light easy drinking style, making it very approachable. Floral in the background, with low astringency.

DRINKABILITY: 7/10

Juice analysis: the science of Dorset cider apples

Variety	Taste	SG*	Brix**	Malic acid (grams/100ml)	Tannin (grams/litre)
Late September					
Lancombe	Bittersharp	1.0435	10.3	1.51	1.8
Hains Sweet	Sweet, sub acid	1.0503	12.3	0.54	1.4
Dewbit	Sharp	1.0383	9.4	0.88	1
Hunter's Ground	Full bittersharp	1.0540	13.4	1.59	5.6
Frome River	Full bittersweet	1.0495	12.2	0.20	2.4
Symes Seedling	Sweet, sub acid	1.0441	10.7	0.67	1.6
Meadow Cottage	Sharp with tannin	1.0472	11.3	1.85	1.9
Early October					
Rough 'n' Reddy	Full bittersharp	1.0517	12.5	1.63	7.9
Tom Legg	Sweet	1.0426	10.8	0.22	1.1
Cattistock Pink	Full bittersweet	1.0521	12.8	0.32	2.1
Buttery Door	Sharp	1.0507	12.2	0.92	1.3
Marnhull Mill	Full bittersharp	1.0488	12.4	1.90	10.6
Mid October					
Matravers	Full bittersweet	1.0480	11.6	0.19	2.8
Puddletown	Full bittersweet	1.0430	11	0.31	2.7
Tangy	Full bittersharp	1.0413	10	1.18	2.6
Winter Stubbard	Medium bittersweet	1.0507	12.7	0.30	1.6
November					
Golly Knapp	Full bittersweet	1.0517	12.7	0.29	2.40
Marlpits Late	Medium bittersweet	1.0404	10.0	0.20	1.50
Golden Ball	Sharp	1.0417	10.2	0.50	1.50
Yaffle	Medium bittersweet	1.0454	11.3	0.19	1.40

* Specific Gravity: the weight of a volume of fluid or solution as compared to the weight of the same volume of water.

** Measure of the dissolved solids in a liquid (°Bx).

covered with an inch or two of soil and protected from pests under a layer of thorn or furze. The strongest of the germinated seedlings would be used as rootstocks, but there would always be a few weaker seedlings left behind that might grow on to produce good looking apples. Those that had inherited the right flavour would qualify to join the other cider trees in the orchard. Some of our Dorset cider apple varieties would most probably have been selected this way, such as the excellent Harry Masters Jersey. In Mortimer's time, those selected would have been crab-like and *bittersharp*.

Our Dorset apple finds have turned out to be a very disparate band, all with distinct characters and all suitable for making cider, whether used alone as a single variety or blended with others. Many of them are very different from the kinds of cider apple found in other counties, but the best are bittersweets selected and 'civilised' by selection or just plain serendipity.

The finest of the Dorset bittersweets we found is Frome River, well balanced though full of soft tannin. The medium bittersweets, Matravers and Puddletown, follow closely. All are true to the best traditional bittersweet standards. Golly Knap, probably a seedling from the farm orchard, is another full bittersweet. It is really too tannic to be considered on its own but is without doubt in the same league. All these apple varieties will no doubt have the same ancestors as many of our familiar cider apples, such as Yarlington Mill, Dabinett and Somerset Redstreak, and probably originated as seedlings from good quality pomace.

Easy-drinking Tom Legg and Golden Bittersweet have very little sharpness or tannin. Their sweetness is a more 'Devon farmhouse' character. Yaffle, another pure sweet, has inherited the curious woody, slight mustiness associated with the Devon varieties Northwood, Sweet Woodbine and Slack-ma-Girdle. Winter Stubbard, rated best single variety in our cider tasting, has a pure and clean flavour with just enough smooth soft tannin to make an excellent cider on its own. The ancestors of this sweet group probably came from East Devon and the western borders of Dorset, among numerous varieties with a similar taste that are still preferred by many cider traditionalists in Devon.

There are three apple families with sharp tasting fruit, but the best flavours of all come from the 'true cider sharps'. Kings Favourite has a good flavour and, like Tom Putt, will make an acceptable single varietal. Whereas Lancombe is one of the sharper ones but only proves useful blended together with bittersweets, like

Crab apple in full splendour, Powerstock, Dorset.

the two we found it growing with. Golden Ball, however, is clearly the champion of Dorset sharps: a true sharp cider apple with all the prize-winning attributes to make it an excellent choice for any orchard designed to produce that traditional Dorset cider flavour.

Most traditional farm orchards have a motley collection of cooking and *dual purpose* apples, generally quite sharp varieties that are useful in the kitchen, as well as the cider house. Buttery Door and the ancient varieties, Stubbard, Dewbit and Ironsides are in this family. Their sharpness tends to be culinary in character, since they derive from the ancient *codlins* and early cookers dating back to the sixteenth century.

The three worthy eating apples that we found are the early ripening Symes Seedling, Hains Sweet and the ancient Stubbard, mentioned above. The first two were clearly raised relatively recently for eating, but both make surprisingly good, easy-to-drink cider with a popular modern taste. Again, these have evolved from the same general eating and cooking apple genetic pool.

The bittersharps are the oldest family, derived from wild crab

View across the Marshwood Vale to the sea.

apples, and should perhaps be called the traditional cider apple family. You may have heard of the legendary Kingston Black, even the excellent Stoke Red – these are bittersharp apples which, in a good year, can produce a cider with a pleasing balance of bitterness, acidity and sweetness, alone and without blending. Several of our Dorset finds are bittersharps with moderate levels of both malic acid and tannins, but none has a 'balanced' juice. Most are either too sharp or too tannic, and some are positively over endowed with both. These are the feral crabs, atavistic, holding on to primitive wild genes. Some are natural *wildings*, children of a wild crab mother tree visited by a roving bee with tame domestic apple pollen on its legs. Some trees that we found could be offspring from the traditional practice of rootstock production from wildling apple seed – this is the character of the little trees we found growing serendipitously in the hedges. Bittersharps, like Mutton Street Marvel, Tangy and Rough 'n' Reddy, are all near enough to wild crabs, but full of character. These all have the taste that, in his time, John Mortimer would have selected.

Owing to the unique history and geography of Dorset, the Linden Lea collection, with its extraordinary range of flavours, reveals the changing taste of cider throughout the centuries. Many of these apples undoubtedly played their part in the award-winning ciders made in the villages around Bridport in the early part of the last century. Now that their future is secure, they will continue to add their distinctive flavours to Dorset's *modern ciders* in years to come.

There is so much potential in these rediscovered varieties. Nick and I hope that all our work over the last thirteen years, culminating in the planting of Linden Lea and our 2020 taste test, will really benefit Dorset's cider community. A major planting of these old varieties in new orchards would give craft and commercial makers a unique product to take to market. It will be the next generation of cidermakers who will then be able to claim that their ciders are produced from genuine, unique Dorset cider apples.

Is Dorset the birthplace of English cider?

Although it all started as a hunt for old apple trees, so we could make some cider with a traditional Dorset flavour, our quest has grown and revealed a few very unexpected secrets. Our apple hunt took us all over the countryside, happily searching derelict orchards, hedges and back gardens to collect wood for propagating our own orchard. After a considerable amount of detective work, including apple DNA genotyping, we are beginning to wonder if our corner of England was the first place where cider was made in these islands. Many, many centuries ago, when our Celtic forebears on both sides of the Channel were settling down to enjoy farming life, perhaps the story of cider also began.

Windfalls, Powerstock, Dorset, *c.*1989, by James Ravilious.

A Pomona of Lost and Found Apples
with stories of their discovery

———————————

Stalbridge

Trent

Sherborne

Liberty Fields Orchard

Melbury

Beaminster

Marshwood

Netherbury

Cattistock

Lambert's Castle

Shave Cross

Linden Lea Orchard

Frome Vauchurch

Melplash

Maiden Newton

Whitchurch Canonicorum

West Milton

Powerstock

Wynford Eagle

Pymore

Loders

Nettlecombe

Puddletown

Chideock

Askerswell

Hardy's Cottage

Charmouth

Uploders

Lyme Regis

Shedbush Farm

Bridport

Shipton Gorge

Dorchester

Burton Bradstock

Pucknowle

Weymouth

DORSET

A map of lost and rediscovered apple varieties

Gillingham

Shaftesbury

ton Magna

Marnhull

arminster Newton

Shillingstone

Tarrant Hinton

Blandford Forum

Kingston Lacy

Wimborne Minster

Poole

Wareham

ermoigne

Purbeck Cider Orchard   Studland

Hartland

Worth Matravers

● Location of apple discovery
● Orchard planted with discovered Dorset varieties
▨ Dorset AONB

West Dorset

West Dorset is a land of high windswept hills, green pastures, wooded coombes, and streams winding through meadows and marshes on down to the sea and the cliffs of the Jurassic Coast. The River Axe marks the county's northernmost boundary. Rising from the south side of the river valley is a line of high limestone hills that runs from Lambert's Castle to Pilsdon Pen and on to the spectacular outcrop of Lewesdon, the highest point in Dorset, at 279 metres above the sea. It is easy to see how these landmarks, resisting erosion with the hard, flinty limestone capping, were the naturally made fortresses and refuges for our Celtic forebears in the Bronze and Iron Ages.

The hilly land around them broadens out towards the east and the rough exposed ridge of Eggardon Hill. Further beyond are broad chalk downs and sheep country through which the River Frome flows down towards Dorchester. And beyond those hills is the Blackmoor Vale.

Along the coast between Charmouth and Bridport is another line of hills that stretches from mile-long Stonebarrow in the west to its neighbour Hardown Hill, whose limestone capping built so many West Dorset cottages, and on past the famous outcrop of Golden Cap and the rusty cliffs of Seatown and Eype. Further east of Bridport's yellow cliffs are the farmlands of Burton Bradstock, high above the grand curve of Chesil Beach.

Behind these ridges, protected and virtually encircled these hills, are the pasture meadows and wooded valleys of Marshwood Vale through which the River Char makes its way to Charmouth. This grassy, green landscape is different in so many ways from its upland neighbours. The Vale envelopes the parish of Whitchurch Canonicorum, a disparate rural community of farmsteads and cottages. The village itself is on good, fertile land but surrounding it is difficult farming country with hard-to-manage soil conditions and a complex topography that still gives it a sense of being an impenetrable, wilder country. A good deal of the central part of the Char valley is on heavy clay. Small fields enclosed centuries ago from native forest are connected by wandering lanes between vales and marshes as the land drops sharply towards the sea. Much of the poorly drained wet grassland was unusable until relatively modern times. As the land rises, at first gradually then more steeply over a large part of the eastern side of the vale, it is better drained and more fertile. The pasture here has always been more productive and where there is good drainage, some of the fields will support a little arable. Where the land rises onto the surrounding lighter sandy clay, it is good enough to provide a bit of market gardening and some moderate sized orchards. But Marshwood Vale's greatest agricultural achievement has always been to grow good grass. It is predominantly a pastureland, once famous countrywide for its excellent butter.

There is a broad sweep of rather special soil stretching from Broadwindsor and Beaminster to Bridport and Loders. This is greensand, known as Bridport Sand in West Dorset and as Yeovil Sand in Somerset, where the same soft, sandy, fertile earth lines the valleys of Yeovil and Sherborne. Where it lies below the high chalk areas and in the lower valleys it provides light, easily worked ground for market gardening and fruit growing, although much of it is also excellent pasture. The best of the light soil is also known affectionately as 'foxmould'. It is in this part of West Dorset where all the best cider orchards now lie.

This unique topography and soil composition divides West Dorset into two vastly different areas, each with their uniquely contrasting social history, farming practices and changing prosperity. These differences have had a huge influence on the fortunes of West Dorset's cidermaking, orchards and apple trees. Undoubtedly cider has always been made throughout the whole of the area as an essential part of farming and social life, but the development of how it is made and from what raw materials, has followed very different lines of evolution.

Best Bearer

Golden Bittersweet, Morgan Sweet (Somerset)

Early (late September to early October)
Mild bittersweet cider apple
DNA A3529, match with Morgan Sweet

Large, lightly crowned and pale greenish-yellow, often with conspicuous russet dots. Flowering early to mid-season, but being a triploid it is little use as a pollinator; fruit often drops and rots on the ground and going to waste.

FRUIT: Medium to large; 45–60mm. Flattened conical or cylindrical, broad rounded base, flat nose; broadly ribbed, indistinctly crowned. Stem short, stubby, occasionally projecting slightly from narrow, often heavily russeted cavity. Eye basin small, tight, puckered, sometimes crowned. Eye open or closed, sepals upright, green. Colour pale green, ripening to pale golden yellow. Often a trace of pale orange flush. *Lenticels* conspicuous as small dots. Russet usually only light. Sometimes scabby. Flesh sweet with some mild bitterness, chewy and creamy.

TREES: Vigorous with upright branches and much bare wood. Rather slow to come into full bearing. Can be *scab* susceptible.

ORCHARD WORTHINESS: 7/10

Best Bearer is just such an excellent name for an apple tree, or for any fruit tree for that matter. It is descriptive and somehow traditional, redolent of a comfortably productive and reliable variety that makes good cider.

It was clearly a popular variety and frequently planted in the orchards of Netherbury and Stoke Abbott's, especially the late nineteenth century. It is a name peculiarly local to West Dorset. Hubert Warren sent a sample of Best Bearer apples into Long Ashton in 1927. We also knew it by its synonym, Golden Bittersweet, and sent leaf samples off for DNA analysis to register its name. When the results arrived, we were dismayed to find that our Best Bearer or Golden Bittersweet variety was a match with Morgan Sweet, that well-known and rather commonplace (we thought) Somerset sweet cider variety!

Clearly some serious detective work in Long Ashton's records was needed to sort this tangle out. Surprisingly, there were no samples of Golden Bittersweet ever brought to Long Ashton, so it was never in the list of vintage varieties. Morgan Sweet was, and plenty of samples were sent in from both Somerset and Devon. However, Mr Warren of Netherbury and Mr Bowditch of Stoke Abbott sent in Best Bearer apples in 1927. The juice samples are identical, confirming that Best Bearer is indeed Morgan Sweet by its local Dorset name.

This leaves us with a few questions, since we have no written records. It seems that where three counties are in competition for ownership of the same cider apple variety and would like to hang on to their preferred local name – Devon may like to put Golden Bittersweet in their county list of cider apples; Dorset will want it as Best Bearer, a very local West Dorset variety.

But somehow Somerset still 'owns' Morgan Sweet. Although not a lot is known about its earlier history, Morgan Sweet was widely planted in the twentieth century and is very popular in Somerset. The original plantings of Morgan Sweet in North Somerset were intended for 'pot' fruit: a cooker for the domestic apple market. Much of the fruit went to South Wales and was distributed principally by wholesalers Morgan of Cardiff, it is said, to supply the miners with an easily transportable lunch. It is still cherished in North Somerset as a tasty, sweet eating apple that brings back happy childhood memories. But Morgan Sweet cannot be its original name since it was clearly renamed after its association with Morgan of Cardiff. And since there are no written records, I am inclined to go with the evidence that we have: Golden Bittersweet of Devon is the original variety, a typical Devonshire sweet cider apple, and both Morgan Sweet of Somerset and Best Bearer of West Dorset were its new names, given when it was introduced to the new counties, probably in the late nineteenth century. Indeed, to some people of Dorset it is still known as Golden Bittersweet.

As a cider variety, Best Bearer has one or two potential disadvantages that we have learnt from many years of observing Morgan Sweet's behaviour. It often crops biennially and can be prone to scab and *canker*. It is a *triploid*, so it is useless as a pollinator, and it forms a big, strong, often unwieldy tree. For this reason, Morgan Sweet was used as a *stem builder* for standard cider apple trees in Somerset during the first half of the twentieth century. Also, its fruit matures very early, usually in September, and its juice ferments very rapidly. This made it popular on many Somerset farms since it was possible to get it to ferment to an early cider in time for Christmas. But its rapid fermentation also gave it a bad reputation for being unstable as a naturally sweet cider and very susceptible to bacterial disorder. It was largely Long Ashton's experiments in the 1930s, into the use of sulphur dioxide (SO_2) and proper *racking*, that overcame these problems and put the Dorset cidermakers of the time on the right track.

Buttery Door

Buttery Dor, Do, d'Or or Dough

Early (mid-October)
Multipurpose apple
DNA A1305 unique
Registered as a previously lost variety

Good size, rather flattened shape, pale green or golden, and mildly sharp tasting. Flowering late April, early May. Season mid-October.

FRUIT: Medium to large, often > 60mm. Oblate to flattened conical, often lopsided, angular or tending to ribbed. Stem medium to short, woody, within a broad, deep cavity. Eye basin small, deep, rather pinched, irregular, occasionally beaded. Eye sepals short and closed. Skin smooth, waxy. Often considerable russet in eye and stem cavities, occasionally spreading in patches and streaks. Lenticels inconspicuous as small brown or red spots. Colour yellow, bright, acid yellow green when unripe. Usually a little speckled, diffuse brownish red flush. Flesh mild sharp, rather mealy, yellowish.

TREES: Vigorous, probably triploid but seems to be regular cropping.

ORCHARD WORTHINESS: 7/10

I first heard this interesting story of Buttery Door and its travels from Rupert Best of Hincknowle, Melplash. He had been contacted by a member of the Scott Daniel family who had moved to a cottage in Bowood, between Stoke Abbott and Netherbury in 1956. Edie Haynes, a lady who lived in the nearby village of North Bowood, discovered that Mrs Scott Daniel was interested in gardening and particularly in the cultivation of trees. Edie told her about an old and semi-derelict orchard in the parish where there was one particular apple which she said was called Buttery Door. Mrs Scott Daniel's uncle Fred Thirlby, a great enthusiast, took a cutting from the old tree when on a visit to the family in the late 1950s. He grafted it onto one of his own trees back home in Romsey, Hampshire. Years later, Mrs Scott Daniel took a cutting from Uncle Fred's tree and brought it back to Bowood. We were given a cutting of that repatriated tree from which we were able to propagate many more Buttery Door.

When Buttery Door cider was tried at Long Ashton in 1926, it was described as 'fair, with an aromatic flavour but fermentation too fast'. Sometime later a sample of fruit from Dorset was exhibited at the RHS Crystal Palace Show in 1934, listed as an early sharp cider apple. It is also in the National Fruit Register with several spellings of its name, most of which I suspect are just phonetic interpretations, such as Buttery Dough, Dor or D'Or. There has even been speculation that D'Or could mean it was French and that perhaps this was one of the apples that the monks from Normandy brought to this country. Although Edie Haynes had probably never seen the name written down, she gives it as Buttery Door. The general consensus of opinion has decided that this very old English sounding moniker suits the variety best.

The apples on our young trees at Linden Lea are a good size, rather flattened in shape, pale green or golden with a little russet and mildly sharp tasting. It has been described quite appropriately as a 'pastry apple', one suited to making dumplings. It is sharp, well flavoured and actually cooks well. But as typical farm orchard dual purpose apples, many would also have contributed to Bowood's cidermaking.

Dorset Winter Stubbard

Late season (late October)
Bittersweet cider apple
DNA A1293 unique
Registered as local variety 2019

Knobbly green fruit with indistinct ribs rising to a very bumpy, crowned, pinched and puckered nose. Flowering mid-season. Fruit is not ready until late October but keeps well.

FRUIT: Medium to large, 45– 60mm. Conical with a broad base and a small, pinched nose; often lopsided, with five or ten indistinct ribs. Stem long (15–20mm), projecting distinctly from a deep, heavily gold russeted cavity. Eye basin small, narrow, irregularly five or more crowned. Eye closed, sepals short and downy. Colour Bramley apple green, usually without a flush but occasionally a trace of yellow on one side. Skin smooth, slightly waxy, occasionally with spreading patches of russet. Lenticels may be black or red dots on cheek. Flesh mildly sharp, rather culinary, sweet with slight astringency; chewy, greenish, browning rapidly.

TREES: Spreading, with fruit closely clustered on weeping branches. Possibly triploid. Cropping rather biennial.

ORCHARD WORTHINESS: 6/10

Stubbard

Michaelmas Stubbard, Summer Stibbert, Summer Queening
Very early season (August to September)
Bittersweet cider apple
DNA A3528 matched Summer Stibbert

Fruits are unmistakeable, large and knobbly lemon yellow with a pinched and furrowed nose. Ready very early in August to September.

FRUIT: Medium to large, 45mm to over 60mm. Rounded cylindrical with a pinched nose and small rounded base; heavily ribbed from base to apex, irregular, angular and often lop-sided. Stem thick, fleshy, often strigged, projecting distinctly from a narrow deep cavity. Eye basin slight, shallow, irregular, furrowed and often beaded, sepals reflexed, green. Eye open, or closed if sepals intact. Skin smooth, dry with a few large russet dots. Pale yellow ripening to butter yellow. No flush or very slight blush of pinkish-brown. Core very large, open; tube a broad cone often extending deeply toward the core. Flesh distinctive flavour, sharp; white or yellowish; melting and juicy.

TREES: Strong and spreading as they mature. Fairly regular cropping.

ORCHARD WORTHINESS: 6/10

Summer ripening (Michaelmas) Stubbard and Winter Stubbard are closely related, brother and sister even. Both are very old farmhouse orchard apples, not true cider apples perhaps, but nonetheless they make interesting cider. They are West Country dual purpose codlins, large, yellow-green knobbly and ribbed. Their trees are strong-growing, long-lived and trouble-free and they are both *pitchers*, whose antiquity is verified by their ability to root from cuttings. Stubbard apples drop in the orchard ready for the kitchen in mid-August, whilst her brother Winter Stubbard's fruit hangs on the trees until November.

Michaelmas Stubbard is its full name. Usually it is just plain Stubbard, an eating apple especially popular in the South West, that has probably been around in orchards and gardens since the sixteenth century, although its name wasn't recorded by the RHS until 1805.

We found our first Stubbard tree one August day in Mr Warren's old orchard in Netherbury. By then, most of the trees there were relatively young Yarlington, with one or two ancient Golden Ball, but there on its own near the house, was a Stubbard. The apples had already fallen and were carpeting the ground beneath the tree in a sweet aroma of incipient fermentation. A bramble thicket had recently been cleared from its base, leaving a few long strands hanging from the tall, upright, bared branches that supported a dark green, umbrella-shaped canopy.

After that day we found plenty more Stubbard trees, mostly in and around West Milton, Powerstock and Loders. Its earliness and distinctive flavour earned it popularity as a garden apple in the past. I would grow one now, if I had room for another tree. Michaelmas Stubbard comes ready at the same time as Beauty of Bath and, if not quite so good-looking, is certainly more useful and more versatile in the kitchen.

Winter Stubbard looks very similar to its sister Stubbard, apples like caricatures of old country folk, with knobbly pinched and puckered noses. Nick took me to Bonscombe Farm, between Shipton Gorge and Burton Bradstock. Behind a dry-stone wall that snaked up the steep north-facing hillside, was an old orchard with a handful of surviving, mostly dessert and culinary trees. But up by the farmhouse was a tall tree standing alone by the gate, with the farm's name nailed to its trunk. No fruit had dropped, so Nick went back later in October to collect enough to make some cider. He was able to identify it straight away as Winter Stubbard by the character of the apples.

Its cider is mildly bittersweet, a little pale but with a well defined spicy aroma, slightly blackcurrant, and moderately astringent with restrained acidity. It is clearly not the very sharp Winter Stubbard from Devon described by Professor Barker (the first director of the National Fruit and Cider Institute) in his 1937 list of vintage cider apples, so we were not surprised to find that its DNA is unique: another unknown variety. Since it does look very much like a typical Devon Winter Stubbard – a rather weeping tree, with large green fruit – we have now registered our tree as 'Dorset Winter Stubbard'.

Golden Ball
Neverblight, Polly, Go Boyn
Mid-season (mid-October to early November)
Sharp cider apple
DNA A1299 unique
Registered as local cultivar 2019

A small, hard, conical, russeted apple, deceptively green-brown at first, ripening to golden yellow. Flowering mid-May.

FRUIT: Medium or smallish; 40–55mm occasionally larger than 55mm. Shape conical, flat nose, rounded, regular with a trace of soft rounded ribs. Stem thick and fleshy, usually just projecting slightly from a small, narrow cavity. Eye basin small, slightly bumpy, tending to crowned. Sepals short and closed. Skin dry with russet dots, patches often spreading as a fine network across cheek. Colour green-brown at first ripening through yellow-green to golden yellow. Usually a little diffuse orange or pinkish flush, more noticeable as reddish-brown on immature fruits. Flesh yellowish, chewy, mildly sharp.

TREES: Medium, robust, scab-free. Upright, then spreading. Quick cropping from mid-October, large fruit and heavy crops.

ORCHARD WORTHINESS: 8/10

It was back in 1996 when I saw my first Golden Ball tree. It was in the late Hubert Warren's old orchard, behind his cider barn at Hatchlands, Netherbury. He showed me around the trees one September afternoon.

The orchard must have been wonderful in its heyday when the cider barn was in full production, but when I visited it was already in decline. There were many gaps between the trees and plenty of replacements, mostly Yarlingtons, but he took me to his last remaining Golden Ball. I expressed surprise that the apples were dark red, not golden as I'd expected, but he explained that the fruit are late to mature and don't gain their glowing golden colour until the end of October. This autumn deception led us astray more than once in our tree hunt; we would come across 'dark red apples' that seemed to fit no description, only to realise later that we had found yet another Golden Ball.

Golden Ball is certainly the best of Dorset's cider apples. It is a first-rate, medium sharp, low tannin variety, and its name crops up frequently in the scant cider records, confirming it has consistently been a popular choice that suited the taste of local customers. Its juice must always have made a useful contribution to the traditional character of Dorset cider.

Fortunately, there are still plenty of old trees around Beaminster and Bridport, even a few tucked away in odd corners of Marnhull, Kington Magna and around Sherborne. It seems to be a long-lived, lively and still popular variety.

Golden Ball is a cultivar whose origin is unknown, but it dates back from before the great RHS Exhibition of 1883, when it was displayed among hundreds of different apples. There is an earlier but scanty record from Devon in 1830 and its name also appears in the Supplementary List of apples in 'The Orchardist', Scott of Merriott's catalogue of 1872.

Fortunately, many samples of Golden Ball apples did come to the Cider Institute between 1926–33, mostly from the Netherbury and Stoke Abbott area where it was also known as Neverblight. A sample came from a Mr H. Green of Piddletrenthide, Dorchester with the local name of 'Polly'. Many of those apples would have come from mature trees that had been planted towards the end of the nineteenth century. Certainly, there were several Golden Ball trees in the old orchards at Hincknowle before 1898. It was a variety that Mr Pickford from Long Ashton had encountered there during the visits he made for his cidermaking courses and was well known locally as a traditional sharp cider. From his observations he was able to give a brief description of the fruit in the list he made in the late 1930s. The Ministry of Agriculture Bulletin 104, *Cider Apple Production, Varietal Characters of Cider Apples* (1937) includes Golden Ball as a useful medium sharp variety and quotes Dorset as its principal centre of production.

Golden Ball seems to be a robust, scab-free variety that grows strong and healthy, hence its synonym Neverblight. It makes a medium sized, neat tree, upright at first then spreading as it matures. Our young trees were quick to come into production, but unfortunately it seems they could begin to crop biennially. However, I give the variety 8/10 for orchard worthiness.

(Note: There is a false Golden Ball from Monmouth, an inferior sharp apple not to be confused with our genuine Dorset kind.)

Golly Knapp
Late (mid-October to early November)
Bittersweet cider apple
DNA A1296 unique
Registered as local cultivar in 2019 with no known name

Smallish, conical and russeted green fruit with a touch of orange flush and keeps well. Flowering mid-May.

FRUIT: Medium, 45–55mm, occasionally larger. Conical, sometimes slightly waisted and cylindrical, with small, flat nose and rounded base. Rounded in section, sometimes with indistinct ribs, more prominent on nose. Stem quite long (10–14mm) and woody, projecting distinctly from a narrow, deep, well russeted cavity. Eye basin small, narrow and shallow, rather pinched and moderately crowned. Sepals reflexed, usually closed, green. Skin smooth, dry becoming waxy. Russet often heavy and spreading as a network over half the fruit. Pale green, usually with a trace of diffuse brownish-orange flush. Flesh sweet with some soft astringency, chewy texture, greenish, browning rapidly.

TREES: Rather weak and spreading. They show mild virus symptoms but are free from scab. Cropping should be regular.

ORCHARD WORTHINESS: 6/10

Golly Knapp Farm is just north of Puncknowle village, lying in the quiet rolling landscape between Bridport and Abbotsbury. The farmstead, lying about four miles southeast of Loders, was one of many small farms in that hilly area that had its own small orchard in the last half of the nineteenth century. There is little left of the orchard now, other than a few mixed fruit trees – a Kingston Black and a Kings Favourite – but we had come to see the seedling tree that had established itself long ago in the garden.

It was brought to Powerstock apple pressing day, in 2007, and was immediately named Golly Knapp. The tree was in poor condition but gave us enough fruit to make a very nice bittersweet cider with a pleasant hay-like aroma.

Our search for its relations amongst other DNA results has proved puzzling. There are a number of possible close cousins, including several Somerset and Devon cider apples, such as Sweet Coppin, Fillbarrel and Dabinett. Since Golly Knapp is clearly a progeny of the trees that once grew in the old farm orchard, these possible relatives suggest the orchard may have included a good selection of well known vintage cider varieties for its workers to enjoy.

Kings Favourite

Crimson King, Bewley Down Pippin, John Toucher

Late mid-season (late-October)
Bittersweet cider apple
DNA A1302 verified 2017 match with NFC Crimson King
Registered as a synonym

Large, sharp tasting Bramley-sized fruit with a bright red flush and deep open eye. A triploid, flowering mid-season.

FRUIT: Medium or large; 55mm, often more than 60mm. Shape variable, usually conical, rarely cylindrical, broad flat nose, broad rounded base; rounded ribs, often irregular. Stem stout, fleshy or woody, usually level with the base or projecting very slightly; stem basin variable, small, usually quite shallow. Eye basin small, shallow, usually smooth, often irregular and slightly crowned; calyx open, sepals long if not broken, reflexed and free. Skin smooth, waxy, greasy when ripe; occasional streaks of russet, often rough and scaly, green or golden in the stem basin. Scab susceptible, slight *mildew*. Coloured greenish-yellow usually 50–75% flushed with strong dark red, short stripes over bright crimson. Flesh sharp with no astringency; white or greenish, sometimes slightly reddened.

TREES: Vigorous, a broad spreading canopy with much bare wood. Tiploid variety. Likely to be regular cropping.

ORCHARD WORTHINESS: 6/10

Nick and I first found a Kings Favourite tree in a derelict orchard in Pymore, just north of Bridport. The tree, split through to the base and in two halves, was still green and alive, full of glorious large, scarlet apples. We took cuttings then and there to propagate such a fine survivor. We learned that the apples in the orchard once went to Palmers Brewery, in Bridport, to make cider for the company's pubs.

I knew the variety as Crimson King, its Somerset name, but everyone we spoke to in West Dorset called it Kings Favourite. It was recorded as growing in many local orchards at the beginning of the last century, especially around Netherbury and through Marshwood towards Devon. It seems to be a long-lived variety because during our searches we found plenty of strong-growing healthy trees still cropping well.

Being triploid, it makes a large, spreading tree with many bare branches in its youth, but as the bush trees have matured they have begun to furnish more fruiting spurs and now produce a good crop of handsome, brightly coloured apples.

This variety was said to have been raised by John Toucher of Bewley Down near Chardstock. It was known by either his own name, or as Bewley Down Pippin, until it was adopted by the nursery trade and propagated as Crimson King. It soon became popular and was widely planted in western Somerset, adjacent parts of Devon and more recently throughout Somerset.

This seems to be another variety that is shared by the three counties of Dorset, Somerset and Devon under three different names. In West Dorset it is only known as Kings Favourite. It is Crimson King in the orchards of Somerset and parts Devon. But just recently our DNA studies have raised a question mark over which is really which, for we now have a third name, Fair Maid of Devon. All the relatively recently propagated under these three names look the same, and their DNA tells us that they have an identical genome.

Kings Favourite as it has been known in Dorset for over a century is a big, red-fruited, strong growing, sharp apple that makes above average cider. In a favourable year, it can be good without blending. The West Milton Cider Club made a single variety cider from that Kings Favourite tree in the Pymore orchard, and rated it 7/10: 'a pleasingly aromatic cider with a lingering aftertaste,' a respectable appraisal for a sharp cider apple. With such competition, it is no wonder that the Fair Maids of Devon stole their rival's identity! The old Devon variety may now be lost or still exist tucked away somewhere in a mid-Devon orchard. However, anyone wanting to acquire a true Fair Maid of Devon tree nowadays, will more than likely find that it is a Kings Favourite (or Crimson King).

Lancombe

Early season (mid-late September)
Sharp dual purpose cider apple
DNA A1292 unique
Registered as local cultivar in 2019 with no known name

Smallish, yellow fruit, faintly striped with red, with a good sharp cider taste. Flowering late April early May. Ready mid-late September. Fruit falls quickly needing prompt attention

FRUIT: Medium, 45–55mm, occasionally larger. Usually cylindrical, sometimes conical with small flat nose and rounded base. Tending to ribbed from the eye. Stem projecting distinctly (12–15mm) from a small, narrow, deep cavity. Eye basin small, narrow and shallow. Sometimes rather bumpy, tending to crowned. Eye open, sepals quite long, if not broken, green, often upright. Pale golden yellow, usually flushed 50% or less with bright red flecks or short stripes over crimson-orange diffuse flush. Smooth. Skin dry becoming waxy, even greasy. Russet light. Lenticels spots common. Flesh sharp, cream or greenish, occasionally slightly pink. Very juicy with a good cider taste and chewy texture.

TREES: Small and rather weak with plentiful weeping branches. There were masses of small red fruit in year six and again in 2019, but likely to be biennial.

ORCHARD WORTHINESS: 6/10

Marlpits
Marlpits Late

Late-mid season (late October)
Mild bitterseweet cider apple
DNA A1300 unique
Registered as local cultivar with no known name in 2019

Apples are conical, yellowish-green usually with a pinkish or orange flush and often with some russet around the eye. Both the eye basin and the stem cavity are noticeably small. Flowering mid-May. Ready mid - late October.

FRUIT: Medium to large, 45mm and more than 60mm. Size varies depending on season. Flattened conical, with a broad flat nose and rounded base; roundly ribbed, often lopsided. Stem usually short and within the very small cavity or projecting slightly (3– 6mm). Eye basin usually small, shallow but occasionally deeper, often russeted and sometimes beaded. Eye closed. Skin pale yellow or yellow-green, frequently with a trace of pinkish-red or orange flush. Dry, often rough with russet that spreads across the cheeks in a network. Lenticels sometimes conspicuous brown or red dots. Flesh mild bittersweet, yellowish, chewy texture.

TREES: Very strong, upright, and whippy, initially growing tall with few branches. Fruiting in clusters and close into the trunk. Slow to get cropping they are likely to soon become biennial unless given some initial training to improve the upright, fastigiated tree habit. A light crop in the orchard in year 6, but very good in 2019.

ORCHARD WORTHINESS: 5/10

Yaffle

Late-mid season (late October)
Mild bitterseweet cider apple
DNA A1294 unique
Registered as local cultivar in 2019 with no known name

Large broad fruit, yellow-green with some red flecked flush. Ready mid to late October. Flowering early May.

FRUIT: Fairly large, 50–60mm, occasionally larger than 60mm. Oblate or flattened conical with flat nose and broad rounded base. Roundly ribbed and often lopsided. *King fruits* are quite common. Stem variable. In king fruits it is usually little more than a stub, otherwise projecting distinctly (10–15mm) from a deep russeted cavity. Eye basin narrow, irregular and slightly furrowed. Eye usually more or less closed and sepals upright. May be wide open with stamens showing in king fruits. Russet often considerable around the eye, spreading as a network to stem basin. Skin slightly rough, dry becoming waxy. Yellow-green ripening to pale lemon yellow usually with a trace 1/3 cherry-red flecked or speckled flush, dark brown beneath the russet. Flesh mild bittersweet, juicy, chewy texture and yellowish.

TREES: Specimens at Linden Lea are poor, spindly, whippy and spreading. Leaves sparse. The first crop was scanty in year six however they have cropped regularly if sparsely since.

ORCHARD WORTHINESS: 6/10

At the top of the steep limestone ridge to the north of West Milton is Lancombe Cross. There, behind a farm gate, stands Nick Poole's cider house: the birthplace of the legendary, bottle-fermented Lancombe Rising cider and other feats of Nick's alchemy. Nick's little orchard is a mixture of the best of Dorset varieties with a few vintage Somerset ones, tucked away below the cider house pound on the north slope.

On the sunny side of Lancombe Cross lies Marlpitts Farm, with two cider apple trees prominent in its garden. There were three trees getting near the end of their lives when we first saw them, and we christened them Lancombe, Yaffle and Marlpits. Since then, Yaffle has died but fortunately we were able to cut some graftwood from it just in time.

These are trees with great character and their choice for the farm garden is interesting. Marlpitts Farm appears in the 1871 census as owned by W. M. Chilcott, with a 24-year-old tenant farmer called Malachi Daubney in residence (note the Old French surname). When the farm was sold in 1911, the sales particulars revealed its value as a small farm of 29 acres of rich dairy and some arable land 'with a substantial stone-built residence (the farm is sitting on nearly 10 feet of good top limestone and accompanying pit, hence its name) and a compact range of farm buildings, forming a complete smallholding.' This is a fair description of a typical tenant farm capable of supporting a family, a smallholding replicated throughout West Dorset. Three apple trees would be just right to supply sufficient fruit for the kitchen and cider for the family and any farm workers. What's more, the attributes and flavours of the trio are complementary, supplying fruit and juice throughout the season from early to late, both sweet and sharp, and even with a little tannin. Clearly a well-thought-out selection.

The original Lancombe tree, wild-spreading and cropping profusely, is a very early maturing variety that is ready mid–late September. Its apples are smallish, yellow, faintly striped with red and have a good sharp cider taste. They don't keep well when they have dropped and need to be promptly picked up and used.

Their juice is a good, citric full sharp. Lancombe is a dual purpose apple. Suitable for apple pies! And the cider is clean, slightly fruity but without tannin. Being somewhat acidic, Lancombe is a good source of malic for blending with sweeter apples.

Next in season is Marlpits itself, an ugly upright tree dividing at waist-height into six trunks. The juice of its unappetising looking yellowish-green apples is a good flavoured, mild bittersweet. West Milton Cider Club made a gallon of cider in 2009 with fruit from the old tree. The tasters noted that it was 'clean, pale straw coloured.' It had a slight fizz and there was a faint plum aroma. It was voted a very good single variety cider. A more recent cider made from the fruit of the young bush trees was described as 'pleasant but lacking character' – not an unexpected verdict. There is always a difference between cider made from veteran standard trees and that made from young bush trees carrying their first big crop. It does take quite a few years for bush trees to settle down and gain maturity. In their youth, more effort goes into growing branches and leaves with the sole purpose of producing as much fruit, with as many seeds in as possible, to perpetuate the cause. Their fruit will always be larger and the juice waterier and more diluted. Usually, sugar content and tannins will be disappointingly low but typically the acid levels remain fairly constant throughout the tree's life. Hopefully, there will be some vintage quality ciders coming from our young trees at Lindon Lea in the future.

Yaffle was the last variety to ripen at Marlpitts Farm, ready by mid-to late-October. The name is the West Country nickname for a green woodpecker, for Yaffle's apples are large, yellow-green, with a red-flecked flush. This is a good quality mild bittersweet whose cider has a distinctly strawberry, slightly farmyard aroma. It is moderately astringent with a lingering aftertaste and promises to make a fine quality cider in the future. Fortunately, we were able to propagate plenty of young Yaffle trees before the original tree cropped its last.

This trio of trees, with their complementary juices and flavours, would have made many excellent barrels of farmhouse cider.

Ripening over time, their harvesting would have fitted in with all the other things that would have been happening on a busy dairy farm – it was customary on most farms to pick up cider apples when time allowed, with all the family and workers enlisted for the task: a welcome day away from the usual routine. Often the harvested apples would be stored in sacks left out in the orchard, as they still are on some farms. It was also common practice to store the fruit in a dry shed or barn, usually with the different varieties kept separate until there was enough time to get down to the crushing and pressing. Typically, they would wait for a week or two, sometimes longer, until time could be spared to begin the processing. On Marlpitts Farm, Lancombe would have demanded harvesting and pressing quickly, then straight into the barrels to ferment, ready to be blended with the other ciders later on.

It is impossible to know when these three trees at Marlpitts Farm were raised. At a guess, they are about 100 years old, so were probably planted in the early 1900s. They are unlike any recognised cider apple varieties, but clearly they were known locally as 'varieties' since they were chosen to complement each other and probably also expected to produce the sort of cider to suit the local Dorset taste.

Our DNA study shows that the genetic makeup of all three trees is unique. It has again revealed some interesting relatives. The Marlpits variety is associated with the Michelin and the sweet Doux Normandie DNA groups, both varieties being truly French. Marlpitts' *fastigiate* tree habit is very similar to Michelin's upright, multi-leader shape and its juice, like Doux Normandie, is lacking astringency and tannin.

There does not seem to be any close French connection to Lancombe. With its very different looking, attractive red and yellow striped fruit, it may well have acquired its genes from any brightly coloured local apple, but its sharp flavoured juice was more than likely inherited from a wild crab.

Yaffle is different again. This sweet, 'Devon' type is closely related to Sweet Alford, a variety that was once widely planted throughout the South West. We found many Sweet Alford trees in our search. It is an excellent variety that should still be planted, and would be an excellent choice for craft cider orchards. Our DNA research tells us that both Yaffle and Sweet Alford are distant cousins of Cummy Norman. All are low tannin sweet cider apples, and all are susceptible to getting scab. Little is known of Cummy Norman's origin other than that, despite its name, it is not French. Nor is it from Herefordshire but from the village of Cummy in Radnorshire in Wales. This Welsh connection leaves me open to speculating on a possible Celtic ancestry. Could Cummy Norman's ancestors have arrived by boat from across the Channel? Could it have come over with the Breton immigrants fleeing from the Vikings?

It seems quite likely that, in their time, these three trees at Marlpitts were known and frequently propagated local varieties, with French relations or otherwise. They would have been popular in the Loders area in the mid-nineteenth century and definitely before any of the true Norman varieties were introduced from France to Herefordshire. If the three trees at Marlpitts Farm were planted much later, say after the First World War, there is just a possibility that they could be seedlings grown from pomace pips that came from Loders. But I think that unlikely, since not enough growing time would have elapsed in which to assess suitability of the young trees to their site. This makes a strong case for all three to be true Celtic Dorset varieties!

Matravers

Early to mid-season (mid-late October)

Full bittersweet cider apple
DNA A2242 unique
Registration 2019 as local cultivar with no known name
Distributed 2011 with the working name of Loders M

Trees produce regular crops of medium size flattened conical, usually russeted yellow apples, ready in mid-October. Flowering early May.

FRUIT: Small-medium 45–55mm, occasionally larger. Flattened conical with a broad nose and base. Usually regular and roundly ribbed but occasionally slightly lopsided. Stem small, stubby (3-5mm) usually within a small, deep, narrow, russeted cavity. Eye basin small, shallow, narrow. Fairly smooth but sometimes slightly beaded. Sepals upright, long if not broken. Eye usually open. Skin smooth and dry. Yellow-green, ripening to golden yellow, frequently with up to 1/3 flushed orange or bright red, often speckled red. Russet variable, often patchy or spreading from stem to eye as a network over cheeks, sometimes only in stem basin. Lenticels often show as small brown dots. Flesh mild bittersweet, chewy, greenish or yellowish.

TREES: Similar habit to Golden Ball with a neat compact head. Leaves rather pale coloured. Cropping well and regularly.

ORCHARD WORTHINESS: 7/10

By the gate of Matravers Farm in Uploders stands a neat sentinel tree. It is hard to judge its age. It could be well over 100 years old. Four or five strong, twisted branches arise from its fat trunk, not more than two feet high. The juice of its small orange-yellow fruit is good flavoured bittersweet. We named it Matravers.

Matravers' DNA has proved to be unique, but it holds a few French secrets. It bears some similarities to the DNA of Medaille d'Or, a full bittersweet cider apple from Rouen, the capital city of Normandy. This variety was first imported into this country in the late-nineteenth century by the Woolhope Naturalist Society in Hereford because of its valuable vintage qualities.

We found out after our visit that there was once a little orchard not far away at Upton Manor Farm in Uploders that was planted as a trial for the Long Ashton cider research station in Bristol by Mr F. C. Parslow, the local county instructor in horticulture. Amongst the trees supplied by the Bath and West Society and recorded in the 1911 Long Ashton Report, were a number of currently popular French varieties that had recently been imported with the idea of reviving the flavour, quality and reputation of commercial ciders. Cimetière de Blangy and Rouge de Trèves, together with some English varieties like Foxwhelp, Broad Leaf Norman, Court Royal and an unknown variety, Virgin Mary, were in the selection. Although there is no record of the Virgin Mary's origin it could quite possibly have been a local cider variety that was named after the parish church in Loders, dedicated to St Mary. Cimetière de Blangy is very French. It arose as a seedling in the cemetery at Blangy-le-Chateau in Calvados. Medaille d'Or wasn't on the list of varieties in this trial but since it was frequently planted in trial orchards around that time, it was more than likely included the in the Uploders orchard.

A later reference to the trial orchard alarmingly reported that 'the Virgin Mary succumbed to canker in 1923' and had to be painted with H Emulsion. The report goes on to say 'though this orchard is not ideal, it is far ahead of any other in the county, with the possible exception of one belonging to a private gentleman which was planted without any regard to expense.' A clear, contemporary indication of the general plight of Dorset's cider orchards.

There were no further reports made of the progress of the trial trees after 1923 so it is to be assumed that any professional interest in the orchard was probably short lived. Our recent visit to the farm gave us no clues as to where it could have been although reasonably level ground by the river Asker suggests a likely place. There are no local records or recollections of where it might have been or when it expired. It could have been, anywhere in that location.

But its existence led us to speculate on the origin of our Matravers with its French genes. Could it have arisen as a random seedling that acquired its genes from Cimetière de Blangy of Calvados? Or possibly from Medaille d'Or? Time wise, it is just possible that our Matravers tree could have been a random seedling from that old orchard, or perhaps the offspring of any number of trees with reasonable pedigrees that might already have been growing in the area. Even perhaps from a pip in the pomace after a cidermaking session. Whatever its pedigree, it was good enough to have been selected as worthy and planted there by the gate. Perhaps it is that very local Dorset variety, the Virgin Mary? Could it be descended from those cider apple trees that the Bretons brought with them from Normandy in the eleventh century? We shall never know but Matravers is certainly one of the better Dorset varieties that we found. It makes a slightly spicy, fruity, moderately astringent cider, pleasant and drinkable.

Mutton Street Marvel

Late-season (late October)
Late bittersharp cider apple
DNA A960 unique
Registered 2020 as a local
cultivar with no known name
*Small rounded yellow and red
striped sharp apples Flowering
early May. Ready late October.*

FRUIT: Barrel-shaped cylindrical, with a small, rounded nose and base. Rather boxy, tending to be flat-sided in section. Stem short (15mm) protruding distinctly from a small, shallow, bumpy basin. Eye basin small, shallow, puckered and often slightly beaded. Sepals closed. Skin smooth, slightly waxy. Pale greenish-yellow, more than 50% flushed, with light, bright red stripes. Flesh hard and chewy textured. Sharp with some strong tannin

TREES & ORCHARD WORTHINESS: Too young to assess.

Marshwood village's little elevated church is a 170 metres off the road from Crewkerne to Lyme Regis, close to Lambert's Castle. Beside it is a little lane that plunges steeply down. This is Mutton Street, an important ancient thoroughfare, bordered by small farms and crossed by many smaller lanes that form a network over the centre of Marshwood Vale. Mutton Street is a drove road that winds down across the basin of the Char valley via Shaves Cross and Broadoak, towards the better farming lands below, connecting the farms along the way with the town of Bridport. There are still a few old farm orchards, mostly just remnants (the best are on the lighter sandy clays towards Broadoak).

During the time of the Napoleonic War, there were numerous small cottages along Mutton Street, where many of the workers employed in the hemp and flax industry lived. Each croft had its apple trees, some grown in small orchards, but many planted in the hedges and field corners. This is where Mutton Street Marvel was found some years back by Marshwood Vale cidermaker, Tim Beer. It is a sharp tasting crab apple of an uncertain great age, of no particular parentage and typical of many local cider apples. But it crops faithfully every year and withstands an annual hair cut at hedge flailing time. Like the crofters who planted it, the tree is a survivor.

The Mutton Street Marvel is an apple that Tim raised from a graft, and its origins are quite a story. Let Tim tell it: 'This lane had some 42 properties along its length. One such site can be found below Higher Sminhay, an ancient farm dating back at least 600 years. Just to the side of an old orchard you can plainly see the location of one of the original 42 cottages, long gone but leaving its mark on the land. Here, growing from the hedge, is an old apple tree. It wasn't much to look at, but caught my eye as the apples would cling on well into the New Year, and at one stage they were still there in late February, having survived some massive winter storms. Every year without fail, the hedges were hacked back by tractor and flails – it was a marvel that the tree survived at all, in view of the pounding it was given.

'The owners of the land gave me permission to take some apples for identification, and some graft wood to see if I could get them to grow on. The apples were sent to Brogdale in Kent for ID, and a few weeks later a letter arrived confirming what I already suspected: the apple was an unknown, and therefore mine to name. This was confirmed again in 2017, after being genetically tested. It was definitely unknown.

'So there we have it, the Mutton Street Marvel. A 100% Dorset apple. Time will tell as to the cider it produces.'

I call Higher Sminhay Farm 'Dog Bite Corner'. One afternoon when I went with some friends to take a look at Mutton Street Marvel in fruit, the owner's dog bit my leg, taking a big chunk out of it which bled so profusely that A&E was needed to patch it up.

Symes Seedling

Early (mid-September–early October)
Sub-acid, dual purpose dessert or cider
DNA A1290 unique 2017
Registered 2019 as a local variety with a known name

Apples are medium large, tall cylindrical or flattened conical apple, flecked and striped scarlet over orange base. Flowering late April. Season late September - early October.

FRUIT: Medium to large, 45–60mm. Tall flattened conical/cylindrical, flat nose, narrow base; distinct rounded ribs. Stem projecting slightly or distinctly (5–10mm) from a narrow, deep, often russeted cavity. Eye basin well defined, bumpy or crowned, occasionally beaded. Eye closed but open if sepals broken. Skin smooth, dry or waxy. Scab susceptible. Colour pale green to pale yellow. Flush usually 1/3 or less, flecks and short scarlet stripes over pale orangey red base, stripes stronger at the base and fading towards the eye. Flesh soft, with rather weak dessert taste and melting texture. Pleasant eating!

TREES: Strong and spreading. Their dense canopy will benefit from early branch selection and removal. Occasional slight scab. Regular cropping. A light crop in year six and a full one by year eight.

ORCHARD WORTHINESS: 7/10

I think this apple is a first-class find and, what's more, it's all Dorset's own. Symes Seedling is not a cider apple, although it makes a very acceptable pleasant light cider. One bite of its crisp flesh will confirm that it is a good quality eating apple. By mid-September, its handsome red-flushed fruit are ready for picking. They keep well and will store quite happily until December. Symes Seedling is a variety that is still well worth growing.

There is quite a history to this very West Dorset apple. I first saw it, a solitary large red-blushed yellow fruit, when it was brought to Symondsbury Apple day in 2004. It came from a garden in Townsend, Bradpole, without any details on its label other than its name and address. However, we have found plenty more Symes trees in and around West Dorset during our searches, so we know that it was once a popular variety, quite widely planted during the mid-twentieth century, not just in Dorset but also in the wider commercial fruit world.

I found out a little about Symes Seedling in H. V. Taylor's *The Apples of England* (1947) where he claims that it was raised by a Mr Symes from 'Melberry', Dorset, in about 1920. But the family name seldom crops up in the parish records of any of the Melbury villages south of Yeovil, and not at all in the1900s. However, there are plenty of Symes who lived in and around the parish of Loders, and still do. I have been reliably informed by a present resident, Alan Symes, born in Bradpole, that his branch of the family date back to 1721 in the village of Uploders, possibly even earlier than that. Also, there is a Symes Hill to the north side of Loders, between the village and West Milton, so we think it much more likely that Loders was the birthplace of Symes Seedling.

It is believed that around the start of the twentieth century, David Symes, a market gardener had an orchard and a nursery for raising and supplying trees in New Street Lane, Loders (opposite an old hemp-balling mill on the north banks of the Asker). It is clear from the Parish records in the nineteenth century that there were several branches of the family resident in the village. In 1861, most of the men were listed as 'agricultural labourers', a term which could cover many occupations. It seems that the most likely branch of the family could start from George Symes, who is recorded in the 1841 Parish records as living with his family in Loders Mill. He and his wife had three sons, the youngest being David. He would have been in his late 50s by the turn of the century, and so the right age.

The land near the old railway line, where the nursery was, is clearly marked on the 1889 map of the parish as a long narrow orchard extending north under the railway bridge to Smishops Lane. It is not known if he raised Symes Seedling, but he certainly grew the apple here along with another variety popular in its day: Annie Elizabeth. The orchard holding continues to appear on the maps through to the 1930s, but the 1963 OS map shows that the northernmost part has lost its trees. There are several Symes Seedlings still growing in the grounds of Loders Court that were certainly there when Sir Edward Le Breton's family were resident in the early part of the last century. The estate passed to the Hood family in 1999, custodians who certainly value their trees.

Symes Seedling's DNA seems to be quite closely related to that of Best Bearer, that popular cider variety of West Dorset. It is quite conceivable that it is one of the parents from which a chance seedling arose. Did David Symes raise it purposely as an eating apple for local sales? Subsequently, its popularity grew such that it was displayed in the RHS Exhibition of 1934, where it was deemed to have useful commercial value.

Nick and I found plenty more Symes Seedlings in and around Nettlecombe and West Milton. There is a whole orchard of them in Powerstock. We cut our grafts for propagating the variety from one of its vigorously cropping trees. There was only one fruit sample ever sent to Long Ashton, presumably because it was raised primarily as an eating apple, but the Institute described its cider as 'very useful, sweet, good flavour and aroma.' The apples produced from the 2019 Linden Lea crop gave us a well balanced, fresh, fruity and floral cider. It was rather ironically suggested that it would make a good modern cider.

Perhaps that is what Symes Seedling is made for!

Rough 'n' Reddy

Mid season (early October)
Bittersharp cider apple
DNA A3527 unique
Registered as a seedling in 2020

The original Rough 'n' Reddy tree produces vast quantities of large, striking crimson flushed apples quite early in the season, usually already falling by early October. Note the conspicuous starred lenticels

FRUIT: Medium to large, 55–60mm occasionally more. Cylindrical with a flattened nose and rounded base. Usually round in section with well-rounded ribs. Occasionally slightly lopsided. Stem thickish, projecting distinctly from a medium size cavity (8–12mm). Eye basin small, shallow, usually smooth, very occasionally beaded. Sepals wide open, upright, free and reflexed at the tips. Skin is smooth, dry, slightly waxy and without any russet. Pale orange-yellow, usually almost completely covered with a strong bright crimson red flush. Lenticels conspicuous with a pale surround and occasionally starred. Flesh bittersharp, full of astringency, chewy, white or cream.

TREES: The solitary tree is a fair size and quite healthy. Possibly well over 50 years old and bearing a large, matted head of thin twiggy branches. The re-growth on the shoots that we cut for propagating were extremely spindly by nature.

ORCHARD WORTHINESS: 5-6/10

This is the story that needs to be recorded of a rather special Dorset cider apple tree that neither Nick nor I found. It is a discovery that fits in well with the history of the area.

My good friend James Crowden is given to taking long walks in the country. He set out one autumn day from Winsham and made his way southwards through the winding lanes of Marshwood Vale. He reached the village of Burstock and took the footpath that leads up the hill beside the tiny church. As the track rises steeply up the edge of an arable field, skirting a small copse-bordered quarry, it gets ever steeper and more overgrown with nettles and hogweed. Somehow he missed the lie of the footpath but, forging on upwards through the undergrowth, soon became aware of the perfume of ripe apples. The ground beneath the nettles was carpeted with bright red fruit. James, a renowned cider and perry enthusiast, expert authority on everything about the history of cider and an intrepid cidermaker, had found a lone cider apple tree in the hedge. What serendipity!

He took me up there a few days later with carrier bags concealed in our pockets, slightly nervous that we were scrumping apples on private land. The tree was growing wild in the hedge at the very top. James had thought initially it might be Kingston Black but although superficially similar, it was clearly a seedling, more like a cross between Ten Commandments and a wild crab. We each sampled one of the spectacularly large and handsome red fruits. It was rough! A bittersharp taste with plenty of tannin and bags of sharpness, but it had a wonderful aroma. We agreed it was an exciting cider apple that might just make a single variety cider. Needless to say, we picked up as much fruit as we could carry back down the hill and slunk out of Burstock with our booty. A good hour's scrumping!

Later, the apples were milled by the gang at West Milton Cider Club where Nick oversaw their fermentation into a pleasingly dark pink coloured cider with a rather agreeable flavour. James thought a good name for it would be Rough 'n' Reddy, and so it was christened. And now we know that the owner of the farmland is happy to let us have as many apples as we want, we feel more comfortable about taking some cuttings to propagate a few trees from it.

Although there is an old photo of one of the cottages in Burstock, taken in around 1900, that shows the tree up on the hill, there are no longer any farm orchards in the village. I was puzzled to know how a seedling tree might have arrived in such an isolated spot. On the day that we went back for some cuttings, I spent a little time looking at the inscriptions on the gravestones in the churchyard. The family name Curtis on a headstone rang a bell: it was one of the farmers partaking in a Long Ashton fruit crop recording scheme in the early days.

During the late 1930s, all the fruit that was harvested from selected Kingston Black trees at several farms was weighed and recorded every year to monitor how well the variety performed under varying conditions. Mr C. Curtis had two Kingston Black trees that were planted in 1898, in the family's orchard at Whetham Farm, about a mile away down the valley. Those Kingston Blacks will have long gone but it is quite possible that an apple or a pip from Whetham Farm could have been dropped by a passing bird up there on the hilltop. Kingston Black could just possibly have been one of Rough 'n' Reddy's parents.

Tangy

Early mid-season (late September–early October)
Full bittersharp cider apple
DNA A3531 unique
Registration as a probable seedling pending

Very sharp tasting, small, almost elliptical pale golden apple with a long stem. Flowering late April and ready early mid-season, late September– early October.

FRUIT: Medium 40–55mm. Elliptical with a small, rounded base. Rounded with a trace of ribs. Stem long, less than 25mm. thin, green and projecting distinctly from a narrow, deep basin. Eye basin slight, crowned, sometimes beaded. Calyx usually open, sepals green and often broken. Skin smooth, waxy yellow-green ripening to pale gold. Without flush or russet. Flesh full citric sharp taste, chewy and greenish.

TREES: Strong, upright and free branching with a good shape. Needing a little branch thinning and removal in the early years to prevent it becoming multi-leadered. Leaf rather small and sparse but clean. They carried a full crop in years five and 6.

ORCHARD WORTHINESS: 6/10

Nick spotted a little tree covered with green apples in a hedge by the gateway to Moens Farm in Uploders. It had been trimmed almost to a bush by the flail cutter but its fruit was clean and healthy. They were so similar to a wild apple that we weren't surprised by their taste, a super citric sharpness, but somehow quite flavourful, full bodied and with some tannin. We called our new discovery, Tangy.

The juice of the wild crab apple trees growing in the lanes of West Dorset is full of both malic acid and tannin, far in excess of even the most potent bittersweet cider apple. Being a true species, our British native *Malus sylvestris* seedlings should come true to type. But often wild crabs are pollinated by bees visiting cultivated apples growing nearby, their pips giving rise to offspring with a mixture of shapes and colours. More than likely, this is how our self-sown Tangy came to be.

Interestingly, Sherborne's Museum has a collection of water colour paintings by Diana Ruth Williams, a local artist. They are mostly of wildflowers and plants, and one is titled '*Pyrus malus.* Crab Apple, Uploders' dated 1908, depicting a typical wild crab in excellent clarity and detail.

We found out later that the nature of Tangy's DNA indicates that one of its distant relatives could be Cider Lady's Finger, a very old English cultivar that was said by Long Ashton to have been found in a 100-year-old orchard in 1904. This is an elongated conical green apple, like a taller version of Tangy, and characterised by the 'very distinct and unusual flavour of its cider.' I think that rather suggests the 'crab' origins of both varieties. However Tangy does also seem to have a few other even further distant relatives from Normandy, the four bittersweet heavyweights Reine des Pommes, Muscadet de Dieppe, Medaille d'Or and Reinette Abry.

North, Central and South Dorset

Sherborne, the Puddle valley, Dorchester, Marnhull, Shaftesbury and Gillingham were all, not so long ago, the strongholds of good cider orchards and cidermaking. Sadly, Nick and I found that most had disappeared, leaving just a handful, mostly remnants with little sign of cidermaking activity.

The old Ordnance Survey maps show that the villages around Shaftesbury were all surrounded with orchards, all along the lanes and behind the cottages, as recently as the 1903/4 editions. But things have changed drastically since then to accommodate new housing in the rural areas. The report on Dorset's cider written in the late 1930s by Mr Pickford, Long Ashton's Cider Instructor, mentions that scattered orchards were found around both Gillingham and Shaftesbury. Nick and I, guided by a local apple enthusiast, found a few remnants of cottage orchards along Stour Row to the west of Shaftesbury. Some were grafts growing from rescued trees, others were seedlings – some bittersweet but mostly sharp tasting, many of the wild crab persuasion, quite acid and tannic. However, we were told that the fruit from a few veteran bittersweet trees in a orchard just north of Gillingham was still making a fine drop of cider every year. But on the whole our search of north Dorset was unrewarding.

For a nostalgic reminiscence of old Shaftesbury and Gillingham, I urge you to refer to Alan Stone's excellent *Dorset Cider* book and Rob Mullins' recollections of cidermaking and drinking from way back the 1940s. Cider houses like the Fox and Hounds on St James Street and the Ship Inn (aka Hell's Kitchen), high Bleke Street in Shaftesbury, once such important social meeting places, are no longer.

Dewbit

Early maturing (mid to late September)
Mild sharp cider apple
DNA A1301 unique
Registered as local cultivar with no known name in 2019
Originally distributed in 2011 as Sour Cadbury (D6)

Large, rather flat, pale primrose-yellow fruit with an open eye in a deep basin. Flowering late April. Ready by mid-late September. They rot quickly so need prompt processing.

FRUIT: Large to very large, 55mm to over 60mm. Flattened conical, or oblate, broad, flat nose and base. Rather angularly ribbed, sometimes rather lopsided or irregular. Stem projects distinctly (10–120mm) from large, deep, considerably russeted basin. Eye basin large, deep, irregular, bumpy, tending to crowned. Eye closed or wide open. Sepals upright, green at base. Skin dry becoming waxy, even greasy. A yellow-green colour, ripening to primrose yellow, often with up to 1/3 faintly striped or diffuse pinkish flush on the cheek. Flesh pale yellow, sharp, melting, or mealy and dry when overripe.

TREES: Moderately vigorous. Mature bush trees have a good growth habit but are slightly scab prone. Susceptible to Nimrod spray.

ORCHARD WORTHINESS: 6/10

A sunny day in mid-October, 2008. We were on our way to meet Norman in Kington Magna, near Gillingham. He told us that there had once been an orchard behind his house before it was built in the 1970s. Sadly, there was no trace of Godwins Orchard, but he directed us to another orchard on the opposite side of his lane at Dash Hayes, an old manor house with a walled kitchen garden.

We could already see the tops of a few apple trees over the hedge in a large, ageing orchard with a dozen or so sheep-nibbled trees that might have been planted in the late nineteenth century. The owners were kind enough to give us free rein of the orchard in exchange for picking up a few cooking apples for them. We made a quick sketch of the tree positions, tasted an apple from each of them and hazarded a guess as to their identity.

We first identified what we thought might be Sour Cadbury or as it is sometimes called, Yeovil Sour. It had been described by Long Ashton as yielding 'a medium brisk cider with an attractive character in favourable seasons.' Our variety certainly had a sharp, refreshing taste but many of the pale, rather flat primrose-yellow apples were already lying ripe on the ground which rather conflicted with the late ripening habit reported for the true Yeovil Sour. Our candidate's DNA later proved to be unique, so we registered it in 2019 as 'Hitherto Unknown Local Cultivar' under the name Dewbit, a bit of Dorset dialect meaning 'early breakfast' (courtesy of the Dorset poet, William Barnes).

Our Dewbit seems to be a reasonably long-lived and popular sharp cider apple (I have seen several similar apples from that part of Dorset brought in by various people for identification at past Apple Days at Kingston Maurward near Dorchester). A DNA search for Dewbit's ancestry gave a tentative but rather surprising result, suggesting that it could be a very old variety. But it is important not to jump to conclusions. Better to wait until more advanced methods give us more valuable evidence in the future. Maybe one day we will discover its true name, but for now it is Dewbit, our new Dorset apple.

The first Dewbit trees were distributed in 2011 as Sour Cadbury, before we discovered its uniqueness and renamed it. Also, the tree was susceptible to Nimrod spray in the nursery – it caused its leaves to turn brown and drop prematurely after being sprayed. This is not an uncommon response with some apple varieties, but would not normally be encountered outside an intensive commercial orchard.

That day, we found a few more cider apples in the Dash Hayes orchard. Beyond a row of Golden Ball trees, there was a single Ironsides tree full of dull green, unripe fruit and another, possible cider tree with a single red apple up high out of reach. It was certainly pleasing to find that Golden Ball is also popular in that part of Dorset, close to Gillingham. A good day's apple hunting.

Stalbridge Park House is just on the Dorset side of the Somerset border, several hundreds of acres of land surrounded by a ten-foot high wall. The present owner, recently arrived, asked me to identify his old apple trees and give him some advice on caring for them. The orchard is obviously very old, judging by the massive oak trees and one or two veteran apple trees. There must have been many trees growing there at one time, but now only one in ten are left standing, mostly planted in 30-foot rows on a triangle in the old fashioned way. They were large trees, perhaps older than 100 years old, all heavy with cider flavoured fruit, but there was not a single

A puzzling assortment of sharp tasting apples.

apple that I could put a name to.

I was expecting local Dorset apples, but after tasting the first three or four it became clear that these were 'custom bred', very tannic bittersweets and sharps. All were a similar shape, red flushed or pale coloured fruit so probably all from the same seedling source. Happily, there were two old Tom Putts growing together with several big Kings Favourite trees and a dozen or so young Browns Apple.

Who was it who bred and planted these trees? Why did they choose such powerful characters? Were they cider connoisseurs? Were the trees all seedlings from one cherished cider apple? Who suggested planting a few Browns Apple to take the rough edge off the juice? Some formidable cider must have been made from that fruit. Certainly, it is an orchard open to speculation.

According to Stevenson's report of the *Agriculture of Dorset* (1802), the Vale of Blackmoor was still highly praised for 'Cider Beauties of England and Wales'. He writes much on the current discussions of the best methods of raising orchards, although he does not differentiate between farm orchards and those planted in the grounds of houses for home use. He does observe that, 'In the neighbourhood of Sherborne it is common to mix six bushels of sweet apples with three of sharp, in making cider; and in some places a few crabs are substituted for the rough or bitter apples.'

The Sherborne Terrier map of 1834 describes at least 41 orchards in the town, covering more than 67 acres. Villages like Leigh, Yetminster, Trent and Sandford Orcas were surrounded by 'thousands' of trees and each would have had several cider presses. Thornford alone had 35 acres of orchards. Its horse drawn cider press kept going until the 1920s. Sadly, we found only a vestige of orcharding left by 2007. We talked to people who had worked the land and those who made and drank cider, and they confirmed that most orchards had seen their last days soon after the Second World War. The only trees spared were usually Bramleys.

Marnhull Mill

Early season (mid to late October)
Full bittersharp cider apple
DNA A2239 unique
Registered as local cultivar with no known name in 2019

Conical, pea-green, ripening to primrose-yellow and a rather scabby fruit. Ready mid-late October.

FRUIT: Small, 40–45mm. Conical, slightly waisted with a small nose and small rounded base; rounded in section. Stem projecting distinctly (5–10mm) from quite a narrow, heavily russeted cavity. Eye basin quite broad but shallow and slightly crowned. Eye open, sepals free and reflexed. Skin dry. Scab-susceptible. Yellow-green ripening to primrose yellow, frequently with a trace of pinkish-orange flush. Flesh sharp, melting, yellowish.

TREES: Small with compact rounded heads, but the leaves are small and the growth rather juvenile, weak and scabby. There was a fair crop of small fruit in year 6.

ORCHARD WORTHINESS: 5/10

Meadow Cottage

Mid-season (early October)
Sharp cider apple
DNA A2244 unique
Registered as local cultivar with no known name in 2019

Small, round, conical and green fruit, sometimes with a spotty red flush. Flowering mid to late April. They will start dropping early in October.

FRUIT: Small, 40–45mm. Conical with a small nose and small rounded base; rounded in section with a trace of ribs. Stem long (10–15mm) projecting distinctly from a small, slightly russeted cavity. Eye basin small or just a slight indentation, often with beading. Eye open, sepals short or broken. Skin smooth and slightly waxy. Ripening to butter yellow, frequently with a trace of bright red speckling. Lenticels widely spaced and showing as small red or russet dots. Flesh sharp, chewy but juicy, pale yellowish.

TREES: Strong, large, spreading with whippy growth and a thick, drooping canopy. Some mildew. A good crop in years six and nine but likely to be biennial.

ORCHARD WORTHINESS: 5/10

Marnhull is the largest village in Dorset, made up of a half a dozen hamlets, each with their own character. It is a place that one might expect to have retained a reasonable scattering of orchards, but Nick and I were disappointed to find just two sites within a mile of each other on the northern edge of the village, close to the river Stour in the old hamlet of Hains. There was noone left living in the village who might have been able to tell us of their history or how the trees might have been chosen.

We were directed to the end of Old Mill Lane, almost to the river's edge, where there were two small trees in the scruffy meadow, both with a fair crop of small yellow, very sharp tasting apples. They were clearly planted and seemed healthy enough (if a little scabby) in spite of their neglect in an overgrown bit of grassland, so we felt they had some potential to perform well on a good site and an appropriate rootstock. We named them Marnhull Mill.

Close by was a small orchard belonging to Meadow Cottage. The owner, a newcomer to the village, was pleased that we could put names to the eating apples that were growing there. She directed us to a small field opposite that might have been the remnants of an orchard, but now had a single tree that had already dropped all its small, sharp green apples.

In 2011, we distributed young trees of both these varieties. Now we have had time to assess them as bush trees, growing at Linden Lea, it seems most likely that Marnhull Mill is very close to being a crab apple. It makes quite a small tree with a neat, rounded head but its leaves are small and its growth rather juvenile with weak, thorny shoots. The apple juice is not only sharp tasting but astringent and full of tannin, at the sort of levels expected of a wild crab apple. Understandably, we have not made a single variety cider with it, but no doubt it could be useful in moderation for blending where strong body and character are lacking.

The character of Meadow Cottage's juice is very similar to that of Marnhull Mill's: a rather crabby flavoured bittersweet, perhaps with a little less tannin and more mellow than its neighbour. When we came to taste the juice from our own young trees for cidermaking in 2019, Nick described it as 'mouth puckeringly sharp'. It may be that we had judged its character to be better than it really was on the day that we found the tree, but for now Meadow Cottage will have its place as another useful bittersharp for blending. It makes a poor tree in the young orchard at Linden Lea – strong, spreading and full of whippy growth. It tends to get mildew and seems likely to go biennial. It's not one of our special finds. However, both of these river Stour varieties, Marnhull Mill and Meadow Cottage, could probably be quite typical of the sharp, almost crab-like, cider sorts that have been traditionally grown in that part of Dorset.

Hains Sweet

Mid-season (early October)
Sub-acid dual purpose cider apple
DNA *Pending*
Registered as local cultivar with no known name in 2019
Originally distributed in 2011 as Sour Cadbury

Red flushed and heavily russeted. Ready by early October.

FRUIT: Medium, 45–60mm. Flattened conical, broad flat nose and base. Stem thick, quite long 10–18mm, projecting distinctly. Eye basin small, very narrow and deep. Sepals open, upright. Skin slightly waxy. Russet in eye basin, spreading broadly as a network over cheeks. Colour yellow-green, almost completely covered in a strong bright red flecked and striped flush over diffuse orange. Flesh sweet, sub-acid and flavourable. Yellowish, melting, becoming mealy.

TREES: Small and compact. Comes into cropping quickly and regular;y. May need a strong rootstock.

ORCHARD WORTHINESS: 7/10

Tom Legg

Mid-season (mid-October)
Sweet cider apple
DNA A1298 unique
Registered as local cultivar with no known name in 2019

Pale yellow apple with conical shape and a broad, flat nose.
Flowering early May. Ready early to mid-October.

FRUIT: Medium to large 55mm to over 60mm. Conical tending to waisted with broad flat nose and small rounded base. Tending to angular with ribs extending from the nose. Stem short (5–10mm) level with or projecting slightly from a small, narrow but deep, rather uneven cavity. Eye basin small, narrow, rather pinched, slightly crowned and often beaded. Eye more or less closed, sepals long if not broken. Skin smooth, dry, pale yellow-green ripening to yellow, occasionally a trace of orange flush. Occasional patches of russet. Lenticels small green dots or large and russeted. Flesh sweet, melting, yellowish.

TREES: Vigorous, upright with some bare wood, probably triploid. May require early branch removal. They produced a massive crop in year 6.

ORCHARD WORTHINESS: 7/10

Just a few hundred yards from Marnhull Mill, along a footpath across the field to Hains Lane, is a fine old manor house with two orchards. The main walled garden next to the house is planted with well chosen eating and cooking apple trees. Behind the house was a small area of trees where we found two big old ciders accompanied by a jumble of smaller apple trees, a walnut and a medlar. Again, we were unable to find out anything of their history, although the big trees had clearly once been part of a larger cider orchard.

The first tree was small, bearing a good crop of handsome bright red apples streaked with golden russet. This was an early maturing, possibly dessert variety with a pleasantly flavoured sub-acid juice. But as we would later discovered, it makes a well balanced cider and is worth growing for both purposes. We gave it the name Hains Sweet.*

The second was a tall, vigorous old tree loaded with large yellow, sweet tasting fruit that matched the description of Golden Bittersweet, and we distributed it under that name until its DNA later proved to be unique. After a little more detective work, it became clear that we had found another Dorset variety: Tom Legg. The apples matched the flimsy 1938 Long Ashton description: 'Pale yellow, conical shape with a broad flat nose, ready early to mid October'. We found two similar trees on a farm over 30 miles away at Shaves Cross, near Bridport, West Dorset – the family name Legg is quite common in West Dorset, but there were no clues as to who this particular Tom Legg might have been or how he gave his name to a cider apple. But with a little more detail, we were able to confirm their identity and register them as a unique cider apple variety under their true name.

---

* There was yet another interesting tree at Hains, a bittersweet, as yet only listed as Hains No. 2. Our trees at Linden Lea have been very slow to come into cropping and the apples are very late maturing. We are still waiting for them to produce enough fruit for us to make a cider. Let's hope it keeps its promise to make a good bittersweet cider for us sometime.

Frome River

Mid season (early to mid October)
Bittersweet cider apple
DNA A2243 unique
Registered in 2019 as local cultivar with no known name

Typical bittersweet jerseys, conical red with dark crimson stripes. Flowering early May. Ready early to mid October..

FRUIT: Medium to large, 50–60mm. Conical, small flat nose and broad rounded base. Slightly angular with rounded ribs. Stem thick, fleshy, projecting slightly (10–15mm), often off-set, bulge at base. Stem cavity small, narrow, deep, sometimes very small and tight, often cracked, usually with some russet. Eye basin small, narrow, shallow but well defined usually smooth and occasionally beaded. Eye closed, sepals long, upright if not broken, green. Skin smooth, slightly waxy, often scabby. Variable russet, usually in the stem basin and around the eye, occasionally spreading as streaks. Lenticels usually inconspicuous but occasionally small green or brown dots especially on flush. Colour pale green, always 1/3 to 2/3 flushed and lightly striped dark maroon over bright red. Flesh mild bittersweet, white, chewy, occasionally reddened under the skin and red vascular strands.

TREES: Good natural habit with moderate vigour. Cropping precocious and seems regular. This find looks like a very promising variety, but is possibly scab prone.

ORCHARD WORTHINESS: 8/10

Reinette Obry
Known in France as Reinette Abry

Late season (mid October to early November)
Medium sharp cider apple
DNA A1297 matched A2683; 95560 Baillet-en-France

Greenish-yellow fruit, ripening to butter-yellow. Easy to identify from their distinctive lenticels and russet. Ready in mid-October–early November. Flowering mid-May.

FRUIT: Medium to large, 50–60mm. Oblate or flattened conical, with broad, rounded nose and base; rounded in section, sometimes rather ribbed. Stem short (5–10mm), usually within a small, deep, golden russeted cavity. Eye basin medium, narrow, deep, sometimes faintly crowned. Eye usually open, sepals upright, green. Skin pale greenish-yellow ripening to pale butter yellow, dry. Often covered with a distinctive network of russet, but sometimes without. Lenticels corky, occasionally reddened. Occasionally a trace of pinkish-orange flush on some fruit

TREES: Reinette Obry is a shy bearer, very slow to come into cropping but it can be very long lived. It performs well as a bush tree once it gets going. Young trees have stout upright shoots with very short internodes. Growth is good and mature trees are a good size if rather erect and heavily spurred.

ORCHARD WORTHINESS: 7/10

A couple of interesting finds were added to our collection in the final year of our quest. They were trees that itinerant apple-seeker and cidermaker, Russell Crocker had spotted on his travels in the Frome River valley near Maiden Newton. One day in late October he took us to an old farm orchard, part of the Best family's estate in Frome Vauchurch since the early nineteenth century. On the far side of the river there were half a dozen or so very old apple trees; only a few were bearing fruit, mostly rather nondescript cookers, but one tree laden with fair-sized red-striped apples caught our attention. The apples were a typical jersey shape, red with dark crimson stripes and their bittersweet taste confirmed their heritage. This was one of the most promising flavoured varieties we had sampled.

A couple of miles further, in Wynford Eagle, in another part of the old Best family estate, were three trees laden with light coloured fruit, high up at the far side of a field of inquisitive cows. The apples were crisp and sweet tasting. Thanks to the heifers there was not enough clean fruit left to make a cider sample, but we were able to return a few weeks later to collect some graft wood.

Young trees of both these new finds were distributed in 2013, the red jersey bittersweet as Frome Vauchurch Red and the pale yellow as Cadbury. That was all fine until leaf samples went off to East Malling Research Station for DNA sequencing in 2018, and their results threw up some interesting surprises that overturned our first thoughts.

Disappointingly, the DNA of my hoped-for Cadbury matched that of Reinette Obry, a French juice and cider apple that was introduced into this country in the early 1900s. This dashed my hopes of the rediscovery of a cherished 'lost' variety, but the result was instrumental in revealing the story of Reinette Obry's unknown origins. Thanks to a little help from a French cider apple colleague, Sylvain Drocourt, we confirmed that its true name is Reinette Abry from Normandy.

Frome Vauchurch Red's DNA proved to be unique, and we therefore had the opportunity to give it a name of our choice. Frome Valley River being rather a mouthful. Nick's wife Dawn came up with the excellent alternative, Maiden's Blush – only to find that it had already been given to a rather indifferent eating apple. Our handsome cider apple was eventually registered in 2019 as Frome River. However, since then, more DNA details have emerged that throw doubt on its uniqueness.

Since all our Dorset finds have had their DNA sequenced at East Malling Research Station, I have used their 'Explorer SSR' software to find out as much as I could about their ancestry by comparing their sequences with those of known cultivars. The clever software even allows you to search for possible parents by finding all those with a similar pattern of *genetic markers*. I was more than a little surprised to discover that our Frome River is closely related to Somerset's Burrowhill Early, which could well be one of its parents. Another piece of the parental jigsaw was provided by an old variety called Jimmy Male, sourced from Porter's nursery at East Lambrook in Kingsbury Episcopi. This was given to me by Simon Porter, grandson of the nurseryman who fostered so many good vintage-quality cider apples in the early nineteenth century. Porter's Perfection, an excellent bittersharp, is one of his. By putting the two putative parental DNA sequences together, the pattern of markers matched Frome River's almost completely, or enough to confirm that all three varieties were from the same close family of cider apples and that one of them was almost most certainly a parent.

Frome River may not be a Dorset variety after all. But we had found two vintage quality cider cultivars within a couple of miles of each other that were most probably sourced from Simon Porter's nursery. Judging by the age of the trees, this would have been in the early part of the twentieth century, about the time when there was much interesting work going on at the Long Ashton Cider Institute. The Best family, always in touch with current experimental work, both through the Bath and West Society and Long Ashton, would have been among the first to adopt suitable varieties for the improvement of their orchards. Where better to have obtained them than from Porter's of East Lambrook, in the heart of Somerset ciderland?

Hunters Ground

Early season (mid to late September)
Bittersweet cider apple
DNA A1291 unique
Registered in 2019 as local cultivar with no known name
Distributed under the working name Joanie's Apple

Handsome apples, large, brightly coloured, often lopsided, cylindrical and irregular, with an excellent full bittersharp flavour.

FRUIT: Medium to large or very large 55–60mm. Tall, cylindrical with a broad flat nose and base. Often lopsided and distinctly ribbed in section. Stem usually short, often a stub in king fruit, sometimes projecting distinctly (12–15mm) from small, fairly shallow basin. Occasionally a little green scarf skin. Eye basin small and narrow, crowned, irregular and furrowed. Eye distinctly open, often lopsided. Sepals long, upright and green, especially at the base. Skin smooth and slightly waxy, yellow or pale yellow-green. More than half covered with narrow stripes of bright red over a diffuse flush of orange red. Flesh well flavoured bittersharp, very juicy, cream coloured.

TREES: Starting life with good centre leaders but becoming very upright, compact and multi-leadered. Quick to come into cropping but soon becoming biennial.

ORCHARD WORTHINESS: 5/10

Cattistock Pink

Early maturing (mid to late September)
Mild bittersweet
DNA matches Malus Wisley Crab, *M. niedzwetskiana*

Conical apples flushed all over dark brownish crimson. A striking tree in spring with pink flowers and leaves tinted red. Ready mid-September.

FRUIT: Medium large, 45–60mm. Conical with rounded nose and base, often slightly waisted. Somewhat angular, often distinctly ribbed. Woody stem fairly long (20mm) projecting distinctly from a deep cavity. Eye basin small and narrow, rather irregular, often beaded. Eye usually closed, sepals long and reflexed. Skin smooth, dry, becoming waxy. Almost completely covered with a dark brownish red diffuse flush. Flesh chewy, becoming mealy, reddened throughout but lightest around the vascular strands. Taste mild bittersweet.

TREES: Moderately vigorous, upright and ornamental. Scab prone.

ORCHARD WORTHINESS: 7/10

We struck lucky in Cattistock. We were guided to a pony paddock behind Orchard House in the main street of the village. There, mercifully protected from the ponies by a substantial wooden tree guard, was an old tree laden with fine looking, red-striped fruit. First bite told us that these were rather special cider apples, pretty sharp but with some tannin. Subsequent lab analysis of their juice confirmed that we had found an exceptional, full bittersharp cider variety, probably a little too sharp for making a balanced single variety cider but one with bags of unique character. We gave it the name of Hunters Ground in acknowledgment of the local Cattistock Hunt which is so much involved in the local community and social events.

It transpires that the pony paddock and much of the land behind the houses that line the west side of Cattistock's main street was full of trees until relatively recently, and is still called 'Orchard'. Below it, the river Frome runs along the edge of Miller's Barton field that belonged to the Langford family who owned and worked the mill. Buddy Langford and his mother, Marie, still live in the mill, and he remembers the family's orchard that lay between them and the railway line on the west side of the village. He recalls that, as a lad, he and his chums used to throw cider apples at the trains. He later became an ace bowler for the village cricket team; cider apples had their other uses too. He also talked about flagons of cider and bread and cheese being taken up to the fields during haymaking.

Marie Langford has written a little book, *Cattistock – A Dorset Village 1916–2006*, recounting all sorts of interesting people and activities over the years. She talks warmly of haymaking in her youth. 'Hay-making was vastly different from today . . . the only danger was if one of the cart-horses trod on your toes when being led between the rows of hay for the men to pitch the loose hay on the wagon and take it to the rickmaker. There was always plenty of homemade cider for the men to drink . . . the moon and stars would be shining when they eventually decided to wend a shaky way home. We made our own cider when we had an apple orchard – long since gone – the apples would be loaded on the spring wagon and taken to Mr Dewdney's cider press at Sandhills – used as a roost for some of his hens. It was lovely cider and, as children, we were never banned from helping ourselves from the barrel.'

The ancient Hunters Ground tree is still alive and well in the paddock, although a little thinner now through old age and broken branches. A centenarian, in memory of the village's cider history.

On another September day, after a fruitless search of Cattistock back gardens, we returned empty handed through the churchyard towards the Square. We spotted a small dark green tree laden with large brownish crimson apples behind the fence, almost hidden under the yew trees. It was a splendid sight at the end of an unproductive day, although it was clearly not a cider apple, rather an ornamental variety planted there to embellish the graveyard. The apples we scrumped and tasted sweet with a hint of astringency and the faintly strawberry scented, chewy flesh was liberally stained red under the skin. This was enough to suggest that these apples might make a good drop of cider, as indeed they have proved to do. In a good year, their deep pink juice, flavourable even as a fresh drink, ferments to a clean, mild-bittersweet cider. Sadly, the variety is very prone to scab and in a bad year the handsome fruit are coated with dusty brown spots making the cider rather stale tasting and bitter.

From these Cattistock trees, those grafted and planted in the Linden Lea collection are certainly very ornamental. The show begins on warm spring days in May with a full head of delightful crimson blossom. The leaves develop soon after petal-fall, at first pink-tinted then slowly darkening with green. By September, the russeted red fruit are ripe and ready.

Its DNA tells us that it is an offspring of *Malus niedzwetskiana*, the red form of wild crab *Malus sieversii* of Kazakhstan, and responsible for so much of the red colours in our cultivated domestic and cider apples. The commonly planted form is usually known as *Malus Wisley Crab*, but we decided to name our accession Cattistock Pink, giving a nod to the village's story.

Ironsides

This is the true French Crab which has many synonyms

Very late season (late November)
Dual purpose apple
DNA A2240
Match with French Crab, National Fruit Collection

A late maturing Granny Smith look-alike. Hard, dark-green apples distinguished by their prominent white spotted lenticels. Its greatest asset might be its long-keeping quality. Mature by late November but hard enough to keep indefinitely. Flowering mid-season.

FRUIT: Medium, 45–60mm. Sometimes larger. Round or flattened conical to almost oblate. Rounded nose and base, softly ribbed. Stem projecting distinctly (12–15mm) from medium sized cavity. Eye basin small, quite shallow, rather bumpy with irregular shallow crowns. Eye more or less open. Skin smooth and slightly waxy. Some scarf skin in stem cavity. Lenticels a prominent feature. Large with a pale surround, getting smaller and closer together towards the nose. Colour dark green, occasionally with a trace of lightly speckled or striped red. Flesh rather non descript sharp, green, hard and chewy.

TREES: Strong growing and vigorous.

ORCHARD WORTHINESS: 7/10

Ironsides real name is French Crab, an apple that is said to have first been introduced to England in the late 1700s. But it is so well settled in Dorset that I have a feeling that it may have come to our Jurassic shores even before that. Mr Pickford described Ironsides apples as 'Sharp, late, medium, round, green, very hard texture,' adding in pencil '(Stone Pippin?)'. Ironsides was tried for the first and only time at Long Ashton in 1928, when Mr Smith of Stoke Abbott sent in a sample of fruit. Disappointingly the cider was described as 'Not of high value'.

A search for a description of Ironsides in Robert Hogg's *Fruit Manual* proved negative. Although there were several different Stone Pippins, none seemed appropriate. Hogg had been sent a Dorset apple from a Mr C. T. Hall of Osmington Lodge in Weymouth, called Iron Pin, a small sharp cooking apple which keeps well. The list of apples displayed in the RHS Grand Exhibition in 1934 gives Ironsides as a small round, green, late sharp cider apple from Dorset, but there were no Iron Pins or Stone Pippins.

In order to sort out these names, I paid a visit to the National Fruit Collection in Brogdale, near Faversham in Kent – quite a journey from Somerset but well worth it, if only to see the impressive collection of over 2,000 apple trees, all so varied and spectacular in the huge orchards. There were plenty of cider apples, but no Ironsides. I found two trees of Iron Pin, Mr Hall's Weymouth apple, as hard and green as Hogg had described. There was also a variety called Leather Jacket which seemed identical. This was in October, and they still had a very long way to go to ripen.

A friend of ours found a November maturing apple in a farmyard in Cattistock which fitted the 'Sharp, late, medium, round, green, very hard texture,' description. He called it Cattistock Pea Green. There was another ancient tree near Maiden Newton, with hard green apples, half covered with sooty black fungal mould. These he dubbed, Dirty Maidens. Both finds compared well with Brogdale's Iron Pin apples. Then we found a tree with similar apples growing in Powerstock, and yet another tree in Colyford, just over the border in Devon. Surely these were all Ironsides?

I went back to my books for further detective work, and there under French Crab in the *National Apple Register*, were over 20 synonyms. Way down the list was the name Somerset Stone Pippin along with Ironsides. So, Stone Pippin, Ironsides and Iron Pin were one and the same as French Crab, the original name that it had been given when it arrived in England centuries age. Ironsides and Iron Pin were the familiar names that our West Dorset farmers knew their apples by, but Somerset people knew it as Stone Pippin.

The *National Apple Register* also gives 'John Apples' as a French Crab synonym. It was thought that they were the very John Apple of Shakespeare's time – 'I am withered like an old apple-john' (Henry IV Part I). Edward Bunyard, who was England's foremost pomologist and fruit tree nurseryman in the early twentieth century, not only corroborated this in his *Handbook of Fruits* but was able to add a few more synonyms to French Crab's list, one being Iron King. Several horticultural writers have claimed that French Crab arrived this side of the Channel in the late eighteenth century. This would have been too late for them to be Shakespeare's John Apples, since Shakespeare died in 1616. More than likely, the date of their introduction was not founded on contemporary evidence. However, it is quite possible that our Ironsides did come over from Normandy, incognito, to our West Country shores in the sixteenth century as a cider apple before it made its way northwards into English gardens. We can only speculate. In the past, apple tree names would hardly ever have been written down but learnt by spoken word to be passed through the generations. This gives plenty of opportunities for changes along the way. Names were forgotten, names were invented. It is easy to see how it happens.

We have kept the name Ironsides for our Dorset trees growing at Linden Lea. As a young healthy bush tree, Ironsides produces apples that are reasonably sized, handsome dark-green cannonballs, majestically burnished with silver spots of russet. Edward Bunyard liked it, and goes on to say in his book that it is a valuable late season fruit that cooks excellently. Perhaps not to modern popular taste, but a fine old dual purpose variety.

Liberty Belle: Dorset's Adopted Companion

Early season (mid September)
Low tannin cider apple
DNA A3536 unique
Registered as a known seedling in 2021

Large, rather irregularly ribbed fruit, with a prominent red striped flush. Flowering early May. Ready mid-September

FRUIT: Large, 55mm to over 60mm. Conical or slightly cylindrical. Often slightly lopsided and broadly ribbed. Stem projecting slightly from a small, narrow and deep basin. Eye basin narrow, irregularly crowned and often beaded. Sepals long, green and closed. Skin smooth, waxy and free from russet. Colour butter yellow, always 30–60% covered with a lightly striped bright red flush. Flesh chewy, yellowish and mildly bittersweet.

TREES: Rather weak, sparsely spurred and tip bearing.

ORCHARD WORTHINESS: 8/10

This variety must surely be adopted as one of our Dorset cider apples. The tree came to Liberty Farm in Halstock, just in Dorset, when one of the fields, called Pythagoras, was being planted with apple varieties and perry pears. The seedling was rogue in a batch of Debbie, one of the Long Ashton cider selections. Robert Imlach, one of the owners, wanted to call it Liberty Belle so that it would still be a 'girl', and have reference to their orchard. The name combines a nice play on the American Liberty Bell with the connections between Thomas Hollis – an original owner of Liberty Farm – and his involvement with the American Declaration of Independence. Thomas Hollis (1720–74) was a philosopher and philanthropist. A friend of Canaletto and Pitt the Elder, a Fellow of the Royal Society, he was also associated with Benjamin Franklin and John Adams, and is regarded as one of the architects of American independence. In Dorset, he named all the farms on his inherited estates after either his heroes or his ideals – Liberty Farm, Marvell Farm, Locke Farm, Springfield Farm and Harvard Farm are those that have survived, and the field names also commemorate his interests – Brutus, Cassius, Cicero, and Pythagoras amongst many others.

We hoped that the true origins of this pretty red apple might come to light through the results of sequencing its DNA, but its genome proves to be unique. Liberty Belle is not one of the 'Girls'. It is not a variety amongst the National Fruit Collection. It is as yet an 'unknown' from an unknown source. Liberty Belle is now launched as a rather special juice apple that will form part of the recipe for one of Liberty Farm's outstanding products. Some mistakes do indeed have a good ending.

Puddletown

Mid season (mid October)
Full bittersweet apple cider
DNA A129 unique
Registered as local cultivar with no known name 2019
Distributed 2011 with the working name, Loders P

Uniformly medium sized fruit, flattened conical yellow green and russeted. Ready in mid-October. Flowering late April to early May

FRUIT: Small to medium, 45–55mm, sometimes larger. Flattened conical with a narrow nose and broad rounded base. Usually regular but occasionally lopsided, rather angular. Stem small, stubby (<5mm) within a small deep, narrow, slightly russeted cavity. Eye basin small, shallow, narrow, tending to crowned. Sepals upright, long if not broken. Eye open. Skin dry, yellow to yellow-green, frequently with a trace to 1/3 flushed orange or bright red, often speckled red. Russet variable, often spreading from stem to eye as a network over cheeks. Lenticels often small brown dots. Flesh mild bittersweet, chewy, greenish or yellowish.

TREES: Open and vigorous. Slight canker. The leaves are noticeably small and dark. Cropping precocious; ours produced a heavy crop in year six but may become biennial.

ORCHARD WORTHINESS: 7/10

Nick and I had only a few responses about apples and orchards in the Dorchester area. There was one Sweet Coppin tree but most were uninteresting eating and cooking apples. We later had a call from a keen cidermaker in Puddletown, who told us that not only were there still a number of old apple trees behind the cottages in Mill Street, but that Puddletown had its own cider club. Although most of the garden apples turned out to be cookers, there were two tall trees behind her cottage that looked very different and promising. They both had a generous crop of small yellow-green russeted fruit that looked very similar to the Matravers tree that we first called Loders M. But Puddletown's trees were different. This variety grew a dense head of dark green leaves, not quite the same – very similar but different. We dubbed it Loders P and waited to see how both it and Loders M might develop as young trees in the Linden Lea orchard, and as they began to mature it became clear that they were quite different despite the similarities we had seen in the original old trees. Both were good bittersweet varieties, and both came into cropping quite quickly. But it wasn't until 2019 that the DNA confirmed both as unique cultivars. We then gave Loders P the name Puddletown and called Loders M, Matravers.

It is hard to get any clues to the ancestry of Puddletown from its DNA profile. There is a suggestion that it may have some similarities with Meadow Cottage, the sharp tasting variety we found in Marnhull. We will have to wait a little longer to find its true lineage.

We learned from the Puddletown Cider Club that there was a small orchard in Puddletown's Old Vicarage, mostly cookers, and another not far away in Tolpuddle, but there was no-one left in the village who knew of their history or remembered what any of the trees were called. Even so, we were told that the Puddletown Cider Club made a reasonable cider each year from the apples they collect in the village.

A Brief History of the Apple

_____

Lambert's Castle has its surprises and delights, atop one of West Dorset's flat-topped hills. Look east and you can see in the distance as far as Portland, sometimes hidden, shrouded in sea mist. South are the cliff tops behind Charmouth, Golden Cap and Eype, sheltering Bridport below. The valleys are filled with sea, its colour changing with the sun and clouds.

I arrived at my favourite orchard one late afternoon on a cold, bright sunny day in early May to find it in full bloom. There were 50 or more apple trees of all shapes and sizes capped with pink blossom, some in family groups and some standing alone. Perhaps not an orchard in the strictest sense, but a wild orchard, just as it must have looked for hundreds of years. That day it was at its most handsome, with the bluebells also in full bloom: a vibrant blue carpet to complement the pink. A wild orchard at its best.

The apple trees are self-sown and randomly spaced. Between them are groups of birch and rowan with the occasional stunted oak tree amongst the sparse bracken and heather vegetation. There are few signs of natural regeneration, but perhaps a few seedlings may be hidden amongst the brambles out of reach of grazing cattle. I did find half a dozen freshly germinated apple seedlings on an old cow pat. So they do survive the ruminant digestive system of the resident longhorns. There they were, thrown out in a fit state to germinate and with their very own starter mulch. I hoped they might survive long enough to get past the tasty stage before the cows came back again in high summer. I doubt they did.

The soil at the top of Lambert's Castle is thin, stony greensand. Beneath is heavy clay that forms an impervious layer, making a spring line at the base of the steep, tree-covered slopes that surround the plateau. The vegetation is typical wood pasture, with small groups of trees in grassland and heather that will have provided safe grazing for wild and semi-domesticated sheep and cattle through

Lambert's Castle. At its best, the perfume of crab apple trees in full bloom with the scent of bluebells underfoot.

the ages. Lambert's Castle now belongs to the National Trust. It was once a hill fort on the boundary shared by the two Iron Age tribes, the Dumnonii to the west and the Durotriges to the east of Dorset, who may have shared the refuge in times of trouble. Some of the ancient field boundaries are still faintly visible.

Wild apples

The story of the apple has been told many times: how the melding of the many different species of wild crab apples from China came to a climax in the Heavenly Mountains of Kazakhstan, giving birth to the wild Asian crab, *Malus sieversii*, named after the Russian scientist, Sievers who first recorded it. And how many, many centuries ago this apple travelled west with the transcontinental Celtic traders along the Silk Routes into the Fertile Crescent to mix its genes with the wild crab apples of Asia Minor and Europe, presenting the world with so many options of fruit, from tasty and sweet to sharpest sour and all colours from green and yellow to the darkest red.

It was the bears roaming high up in the Tien Shan mountains that made the first selections by choosing to eat the tastiest, sweetest fruit. Apples pips pass unscathed through mammalian digestive systems. In fact, they benefit from a thorough cleaning before being deposited back on land, ready primed to germinate away from the parent tree. And it was the Celtic traders and their horses who helped those best apples further along the way. For if a horse was on the move with its Celtic caravan, it could be 30 or 40 miles before the pips from its breakfast apples were deposited along the track, together with a good dollop of manure.

There were many trade routes westwards from China and the mountains of Kazakhstan and Uzbekistan, to that Central Asian area that saw the dawning of agriculture in the 'Fertile Crescent' of ancient Persia. The first cultivation of fruits and cereals began there some 5,000 years ago, in the lands south of the Caspian Sea, the countries we now call Iraq, Iran, Afghanistan and Syria. Then the terrain was lush and productive, with adequate water for growing many crops even as far south as Egypt. Many wild fruits were endemic to this part of the world. There were (and still are in some places) natural woods of apples, pears, plums and cherries, and places where wild quinces and apricots grew together with wild grapes. The Silk Road passed through on their way from the Far East to the Mediterranean Sea, busy with traders bringing further horticultural delights. People of many ancient tribes moved into this fertile land to settle.

The fruit trees that the early gardeners grew were sown from seed and would have been quite variable since seedlings grown from pips will always be different from one another, similar to their parents but never identical. The technique of grafting evolved in China, and travelled slowly west to Asia. When this valuable knowledge arrived, the whole process of cultivating fruit trees was transformed. It became possible to propagate plants vegetatively. By transferring a bud or a graft cutting to a rooted stock, each new individual will be identical to its parent. The best individual trees could then be kept and consolidated as *cultivars*, varieties that will always be true to type. Gardens and orchards could have many identical trees and the choicest selections could be passed on from grower to grower. That was some 3,000 years ago.

The apple story from then on is one of appreciation and selection of what we humans thought we liked best. Naturally, favourites were cherished: the biggest, tastiest and sweetest were eagerly collected and swapped. New cultivars travelled widely with people as they moved from place to place. The Mediterranean lands became famous for their cultivated fruits and gardens. Slowly and surely, apples were spread far and wide.

The wild apple trees at Lambert's Castle have characteristically strong trunks, up to 60cm in diameter, perhaps one to two metres high, then multiple branches spreading to form a dense, twiggy, umbrella-shaped head. The largest trees are probably around ten metres high with their lower branches mostly well grazed. It is not an ideal way of growing a commercial orchard.

Europe has its own wild crab apple, *Malus sylvestris*, which is widely distributed throughout most of the continent. Wild crab apple trees still abound in the hedges and hilltops of the West Country. Those often found growing in the lanes of West Dorset are usually rather small and green, sometimes with a hint of pink flush. They are ideal for making crab apple jelly. But their juice is also full of both malic acid and tannins: a rich source of the raw materials for cider!

Some exciting magic must have occurred in the Mediterranean lands when the sweet and tasty Asian apples met with their tart and tannic crab apple cousins. Our native European crab apple, with all its primitive characteristics, has turned out to be hugely important in the development of our present-day apples, not only cider apples but cultivated apples worldwide. In the past much of the credit for the genetic makeup of apples has been given to *Malus sieversii* and its relatives from Asia Minor, but recent laboratory work on sequencing the DNA of thousands of apples has made the surprising revelation that the European Crab is the prime contributor. Our native crabs have had a huge influence along the long evolutionary apple road, adding much character to the basic *Malus domesticus* and the genetic make-up of early cultivated apples. *Malus sylvestris* has sharpened the flavour of our cookers, spiced the character of our eating apples and, above all, added some zest to our cider apples. More than a dash of crab genes is what makes our cider apples a race apart.

These trees at Lambert's Castle may not be the genuine untainted *Malus sylvestris*, our European wild crab apple species. I suspect that over the years some domestic genes have been introduced, since their yellow-green, often pink-flushed fruit does vary from tree to tree and from year to year depending on how heavily they are carrying. Some trees, in a good year, have been known to achieve fruit of dessert size, but usually they will average three or four centimetres. The genuine wild article, *Malus sylvestris*, grows quite commonly in hedges in this part of the West Country. Its small, round yellow-green fruits have the strongest combination of acidity and astringent tannin that any cidermaker could wish for. Few of these hedgerow trees ever get to grow out of shrub size, since they will be regularly flail-cut back into hedge shape. There are a few truly wild trees that have achieved full status growing in isolated spots in West Dorset, notably on Powerstock Common and the sides of Eggerdon Hill. They can be spectacular, growing freely and unpruned, achieving a splendid head of 'specimen' proportions.

The first 'proper' cider apples that our distant ancestors made cider with thousands of years ago were undoubtedly an early modification of the wild crabs. And over the millennia, those crab genes have been stirred about and added to, time and time again. As Thomas Andrew Knight (1718-1839) – Fellow of the Royal Society, first President of the Royal Horticultural Society, scientist, geneticist, and fruit breeder – said of crab apples: 'The native Crab of our woods was first transmuted into a rich Apple by culture through successive generations; and during its progressive changes it became habituated to culture. Like every other plant, under similar circumstances, it grew more and more dependent upon the

Malus sieversii, the Asian Crab Apple. Quite variable, sometimes tasty, often bitter tasting.

care of man, as it became better adapted to his purpose.'

There are several vital elements that our cider apples owe to these crab apples. Firstly, of course, the taste. The juice of the wild European crab, *Malus sylvestris* is stacked full of ferocious tannin and enough malic acid to outstrip a Bramley Seedling. This is strong stuff, far in excess of the sharpest cooker and even more astringent than the most potent bittersweet cider apple. (Typically crab juice will yield 2.0% malic acid and tannins up to 10.0 grams per litre compared with Bramley's 1.20% malic acid and meagre 1.3 g/l tannin.)

Secondly, their texture. Wild crabs are not crisp or crunchy to bite. They are woolly and everlastingly chewy as anyone who tries to eat one will discover. Cidermakers of the old days were clever

Malus sylvestris: Our wild European Crab Apple, seen here growing in a Dorset hedge.

enough to appreciate this wooliness as a virtue, for it boosts apple press-ability and juice extraction enormously.

In Dorset, our wild crab apples trees seem to show some tolerance of, even resistance to, important apple diseases. Although these useful characteristics would have been transferred along the lines of cultivation, the remnants are fading fast from most of our traditional apple varieties, largely through orchard intensification and the use of plant protection sprays. Strong resistance to scab and mildew is now hard to find. In years to come, when chemicals are no longer used and the need for gene editing becomes inevitable, our wild *Malus sylvestris* is going to be an important source of genetic material once more.

Who made the first cider?

There is a legend which claims that the first cider in England was made in Dorset by monks in the monastery at the village of Loders. It is certainly true that those monks did make cider, and were most probably busy making it in Loders well before William the Conqueror's army arrived.

But perhaps it all started here much longer ago? Indeed, when did people discover that apple juice was so much more enjoyable after nature had worked a little magic? Was it our Celtic inhabitants up on Lambert's Castle, way back in pre-history? What were our ancestors doing here in Dorset while the first apples were spreading through the Mediterranean? How did those wild crab hybrid apples from Central Asia come to England? Who brought them? Did the secret of making cider come from the Far East with them? Why is it that Dorset's ancient cider apples are so different?

We may never know the whole story, we can only speculate. But it is incredible that there are now thousands and thousands of different looking and tasting apples for us to choose from. To see how the history and geography of Dorset – and especially West Dorset – have influenced the evolution of our county's particular apples, we need to go back to those very early days of cidermaking.

In Mesolithic Britain, before 4000 BC, small numbers of hunters and fishers scraped a living in the more sheltered southern coastal areas. They understood fire. They had axes and crab apples (which they certainly ate) but no pottery, so nothing to put cider into. Theirs was a world of trees with a succession of grassy chalk hills and ridges.

As time went on, our ancestors cleared the woods and settled in the more fertile lowlands. Up to about 2000 BC, successive waves of Neolithic people crossed the Channel from the European continent to our southern shores, bringing with them their goats, cattle and the primitive Bronze age sheep (part mouflon, part an ancient East Asian species). Did these shepherds also bring apples with them on their long, slow advances across Europe? Probably. We will never know for certain, but they certainly brought the art of pottery and the knowledge of making useful containers for collecting, storing and holding liquids – all you need for fermenting and drinking a little cider. We already had plenty of our own apples, even if they were only the common wild crabs. It is easy to imagine that sometimes apples began fermenting during prolonged storage, a natural recycling process, and the liquid product must have been sampled. Appreciation of the intoxicating effects would surely have encouraged the samplers to repeat and refine the process, then pass on the secrets to their friends and neighbours.

The sceptical will shake their heads and say that it is impossible to get juice out of a wild crab apple, that it is too small and too hard to crush, and that primitive people without the necessary equipment would not have been able to make cider if that was the only raw material they had. I challenge that notion, and my brother met this challenge in the twentieth century at the age of 18. We had plenty of crab apples in our garden at home. My brother asked a local publican how he might make cider from them. In Surrey, where we lived, the local pub still made and sold draught cider, although its consumption was closely restricted to two pints a night for youngsters of my brother's age. The cidermaker's instructions were: 'leave the apples to get soft, doesn't matter if they go brown – put them in a sack and they'll crush fine.' My brother did have the advantage of a good hessian sack and my mother's mangle to do the job. I helped with the vigorous turning of the handle while he fed the sack of **bletted** fruit through the rollers, and the dark juice squeezed through the mesh. Luckily, my brother's juice kindly fermented to cider and not to vinegar.

Our Neolithic forebears could well have waited till their wild crab apples were really ripe and soft, even let them rot a little before pounding them in a grain crushing quern. Or perhaps they beat them first with a flail made of two crab sticks hinged with a thong? Or maybe they trod them like grapes at wine harvest? All those actions would have softened the flesh and even mellowed the taste – bletting and bruising helps oxidise the tannin molecules, reducing their bitterness and astringency. This is a practice that many a young trespasser through the ages will have learned when scrumping in a cider orchard: *bash your bittersweets to make them sweet!* The miraculous discovery of cider surely didn't just happen once. It must have happened many times, discovered with delight by many people, in many places, long ago throughout the apple world, just as the discovery that grapes could make wine was made.

The Atlantic Arc

The Cornish peninsular was renowned for its gold and tin. Such was the value of these metals that there was regular trade between Britain and the Continent long before the Romans arrived. The Phoenicians were our earliest visitors. They were traders and general carriers throughout the ancient world from perhaps 1000–300 BC. Greek and Phoenician coins, even gold torques, have been unearthed in various places near the coast in Dorset, evidence of the early trade with the Durotriges tribespeople. They came by sea, working their cargo from the Mediterranean around the Atlantic coast, calling in at trading posts along the shores of Spain, France and Western Britain. Cider would not have travelled well over those long voyages, but the Phoenicians might well have brought apples with them to eat. And might they have passed on the secret of fermentation to the Celts of the Atlantic Arc?

It was the wandering Celts who were responsible for bringing us the 'new' hybrid apples, gradually emerging from the wild along the trade routes and forests. After all, it was they who gave us the name to the apple, *avall* or *aball*, which is why Glastonbury was once called *Ynys Avallac* or *Avalon* from the abundance of apples growing there, and why Almaty, the motherland of apples, is the capital city of Kazakhstan. The trees they brought weren't just for growing mistletoe for the benefit of the druids. They bore fruit for the pleasure of both eating and fermenting.

Many Bronze and Iron Age settlements have been uncovered here in Dorset, especially along the river valleys like the Yeo in the Sherborne Vale and the Brit that flows from Beaminster to Bridport, both modern-day cider producing areas. Those moving in from the continent of Europe to settle with our Celtic residents here in the south, would have been descendants of the people who first brought those fruits, nuts and horses out of Kazakhstan thousands of years ago. The earliest immigrants were principally traders, travelling widely and across vast distances. The horses they brought with them as pack animals were the Johnny Appleseeds of the ancient world: apple pips passing through their stomachs undigested, ready to be deposited a good few miles along the way with a handy dollop of fertiliser.

Many travellers found the verdant coastal lands of southern England and the shores of northern France and Spain suited their pastoral way of life. Those that stayed in southern England were by no means primitive. They established extensive field systems in the valleys where the land was suitable and grew excellent cereals. They would have cleared land for their goats, sheep and cattle. Others settled in the southern half of the Atlantic Arc in Brittany, the Basque country and the verdant westernmost, mountainous districts of Asturias and Galicia. Living in these regions gave room for greater independence, away from the intensive influence of Mediterranean culture. The rate of cultural change was more relaxed, allowing each area to begin developing its own 'national' characteristic cider. And as time passed *Malus sylvestris,* the native wild crabs of the Atlantic Arc, would have acquired a few 'foreign' genes and desirable taste characteristics inherited from those Asian hybrid crab apples that came with the westward movement of people.

That was the beginning of *real* cidermaking, and why these exceptional Atlantic coastal borderlands, all Ciderlands to this day, are the 'Cradle of Cider'.

The natural evolution of traditional orchards, widely spaced fruit trees with tall trunks and grass for sheep and cattle, developed from ancient pastoral customs into the highly successful dual farming system we know today. I envisage, in those early days, broad areas of wood pasture, much as Lambert's Castle, with its apple trees, rowans and heather looks now, with our rustic ancestors tending their grazing herds under the shelter of trees: apple trees. Two crops in harmony with each other.

Burr knots, trunk bumps with suckers growing from them, often indicate an ancient variety that will grow from cuttings.

By the late Iron Age, there would have been plenty of genetic variety in our apples. We have no way of knowing when the art of grafting reached us. Many of our oldest apple varieties retain the primitive ability to grow from suckers. Such varieties are referred to as pitchers. Their trunks and limbs bear lumps and bumps called burr knots, from which roots will easily grow when cuttings, usually quite large pieces of branch, are 'pitched' into the ground. This capability signifies their antiquity. It would have been possible for our Iron Age farmer to select his favourite pitcher tree and propagate it by that means, perhaps to grow a few apples in the hedges for eating and drinking.

Our old West Country variety, Stubbard, is a pitcher. In our search for old cider apples, we found many Stubbard trees growing in cottage gardens and farm orchards in West Dorset. Its probably originated from 'stub' or 'stump', referring to its ability to root from cuttings. Gennet Moyle is another ancient pitcher that was widely used as a rootstock in the fifteenth century, until better options for stocks were grown. Our modern rootstocks, like MM106, so useful and efficient today for controlling the ultimate size of our fruit trees, retain that natural ability to throw roots from their buried stems. Most of our modern apple varieties have lost their root promoting genes through successive selective breeding for flavour, size and keeping qualities.

European trade routes

Strabo, the much-quoted Greek geographer who was born around 64 BC, wrote of the trade connection between Britain and Europe just before the Roman invasion. He listed Britain's exports as corn, cattle, gold, silver, iron, hides, slaves and clever hunting dogs. The merchandise we had to offer for trade was brought to the shores via the many pre-Roman roads and trackways formed for the packhorse carriers. Trade with Europe had been steadily increasing for the previous 150 years with great advantage to those Celtic tribes whose land lay near to the Channel coast. Those that grew corn were no longer subsistence farmers but were able to make good money in exchanges.

Prior to and during the Roman occupation, much merchandise from the Mediterranean regions, the hub of trading for all manner of goods from far and wide, was carried through Gaul northwards up the Rhone and into it tributaries, then by packhorse across the mountain passes and overland to the next navigable waterway.

Once down river to the coast, goods were transferred to sea going boats. There were four passages regularly used to bring goods across to Britain that sailed from the mouths of the Rhine, the Seine, the Loire and the Gironde. Strabo had said that, from the Celtic coast one could 'put to sea on the ebb tide at nightfall and land on the coast of Britain at about the eighth hour of the following day.' From the southernmost outlets to our far west coast would have taken much longer but he makes it all sound relatively easy, even comparable with our cross-Channel ferries plying to and fro today.

The main rivers of Britain and Europe were navigable much further inland. The mouth of the River Axe is silted up now but until relatively recent times the estuary was broad, extending several miles inland, probably as far as Axminster with the incoming tide and giving generous berth for small vessels in the lower reaches. Even today, on a spring tide and after heavy rain, the Axe rises so fast upstream that it floods the fields in several places from the coast to Forde Abbey. Ancient Bridport, like Axmouth, was one of many natural harbours where the Britons would have exchanged their merchandise. These small ports offered secure landings, too small and well protected for a major surprise attack from an enemy but secure enough for peaceful trading. It is easy to see that with such reasonably easy communication across the Channel, there would have been plenty of opportunities for some of those fruits and plants that the horticulturaly enthusiastic Greeks and Romans were growing to come our southern coasts.

There were also plenty of more local coast to coast comings and goings. Our Celtic warriors regularly set sail across the Channel to help their friends and allies, the Veneti, on the opposite shore in Armorica (the lands of Brittany) in times of war. I am sure both parties would have been well supplied with cider for the expeditions, made with apples grown and harvested from both sides of the Channel. By Strabo's time, our Gallic neighbours would have already been very familiar with the 'new' hybrid apple varieties arriving from the south. They would be well practiced in wine-making and fermenting techniques from the strong

Mediterranean influences brought buy successions of Roman authority. I feel certain that by the time of the Roman occupation of Britain, in the first century AD, the practice of growing semi-domesticated apple trees, the art cidermaking and the custom of drinking it would have been well established in southern Britain as it certainly was in France.

The gift of horticulture

When the Romans arrived in our corner of South West England, they found a well developed Celtic landscape with villages, hamlets and field systems in the river valleys, cattle in the lowlands and sheep up on the downs. Our people were living successful and fairly civilised rural lives, outside of the occasional skirmishes between the Dumnonii and the Morini or Durotriges tribes. The south coast people were in league with their Gaulish counterparts on the other side of the Channel. All reasons perhaps for timely Roman action in quelling and control; to 'compel', according to Tacitus, 'the Britons to wear out and consume their bodies and hands in repairing the roads and clearing the woods.' Most new roads followed the old Iron Age trackways, suitably straightened for easy traffic of troops and wheeled vehicles. The two main West Country routes, the Fosse Way from Axmouth to Lincoln and the Icknield Way from Marlborough to Norfolk (described at the time as 'the end of the world'), gave free access for straightforward travel and trading countrywide, so important for the spread of 'new' apple trees and plants throughout the land.

Once life had settled down after the invasion, the Romans fully exploited our cereal-growing expertise. In some eastern parts of Dorset they organised large-scale field systems and built granaries, with the tax collector on the spot. As Roman and British peoples became more integrated, Romano-British settlements grew in river valleys, the Yeo vale of Sherborne and along the river Frome around Dorchester. But West Dorset's geography was largely unsuitable for such grandiose agricultural systems. In the main our people were left to work the land in their ancestral fashion, and little attempt was made to convert them to Roman agricultural and horticultural methods.

This was perhaps a time when, in a relatively relaxed world, many new kinds of apple, pear and other more exotic fruit were brought over from the lands around the Mediterranean to fill the luxurious gardens, vineyards and orchards of the villas. The fine Roman agricultural writers, Pliny and Cato wrote of many apple varieties, mostly named after other famous Romans. Those trees will be long gone but there are bound to be traces of their existence left in our apple genetic pool.

The Romans brought nearly four centuries of relative peace between the British tribes. They helped us improve our roads and brought us many things, including rabbits, hot baths and all sorts of novel fruit, flowers and vegetables, and they showed us how to construct villas with pretty, productive walled gardens. But above all they introduced horticulture us to. With their foreign fruit and trees came the ancient Chinese crafts of *budding*, grafting and *pleaching*. Armed with this knowledge, any adept gardener or fruit grower could take a piece of his neighbour's tree and have one of his own, or even profit from selling copies. It was the dawn of nursery practice in this country. Thanks to the vastly improved road systems, many novel plants and some of the valuable horticultural technology and orcharding skills were spread throughout the country.

Alfred the Great and Saxon Wessex

The Romans may have left in 410 AD, but it was probably not for another 250 years that Dorset came fully under Saxon rule, and in the backwater of south-western Dorset, it may have taken even longer. Most likely, there were only a few Saxon settlers amongst the Romano-British people who occupied the countryside and laid the foundations of a medieval pastoral landscape. Although Dorset was part of Alfred the Great's broad kingdom of Wessex, stretching from Kent westwards through Somerset, Devon and Cornwall, there wasn't much to write about Dorset in the *Anglo-Saxon Chronicle*. Our country people maintained their old farming way of life and most probably retained some of their older beliefs.

Because of the first Saxon King, Aethelbert, the people of our island were mostly Christian by the year 700. Orchards and gardens of fruit and flowers bloomed throughout Europe during the ninth century, largely through the enthusiasm and leadership of Charlemagne, Emperor of the Romans and Charles I of the Frankish Celts. He encouraged the planting of vineyards and orchards throughout his empire and set his monks to copying ancient manuscripts, especially writings of a horticultural or medicinal nature. Egbert, who became King of Wessex in 802, had been a refugee in the court of Charlemagne and learned much through their friendship. His grandson, Alfred the Great (871-900), also spent considerable time in France and Italy. Both men would have seen orchards and vineyards on their travels.

Alfred formed a close alliance with the Church and did much to restore the monasteries. One of the earliest references to cider in Europe was made by Charlemagne at the beginning of the ninth century, suggesting that it was also made in a number of English monasteries: certainly Canterbury and Glastonbury, perhaps also Sherborne and Dorchester too.

But it was largely Dorset's geography that discouraged settlement and consequently, a great deal of valuable northern European influence would have passed Dorset by. Most Saxon immigrants, having landed somewhere around Southampton or Portsmouth during the fifth century, either settled in south-east and central England or headed north towards Hampshire and Wiltshire. Here they would have met compatriots moving up the Thames valley. They occupied much of the cereal growing downland and villages, imposing their rule and way of life, taking Charlemage's ideals with them as they went. In the sixth century, further movement west and southwards took them to Gloucestershire and down into Somerset. Dorset was somewhat protected from incursions along its eastern edge by heathland and the dense New Forest. Also,

Wood pasture at Lambert's Castle: shrubs, grass and grazing animals. It suggests much of West Dorset would have looked in ancient times.

during the Roman occupation, a great dyke had been constructed from Cranborne Chase southwards to block the main Dorchester to Salisbury road from invasions. This was still in place years later, but around 650 AD the Saxons finally broke through Bokerley Dyke and reached Dorchester, where one of their warriors is buried at Maiden Castle. This split the Romano-British inhabitants into two groups. The beleaguered western group were set upon by Saxons moving west from Dorchester and more squads that came south from Somerset. The final defeat came at the battle of Penn which is thought to have been at Pinhoe in Exeter, but there is little or nothing recorded of what happened to the encircled inhabitants quietly farming in deepest southwest Dorset.

Those people who settled in central and eastern Dorset were mainly cereal farmers and probably more interested in beer-drinking than growing apples for cider. Nor is the land suited to fruit tree growing. Many were happy to settle in the rich clay valley of Wareham to grow their barley or keep their sheep on the good limestone of the Purbeck hills, both areas that were already well populated by the Romano-British people. All were happy to let their cattle clear the forest of Cranborne Chase. West Dorset was less inviting territory, scattered with small steep-sided hills among heavy, marshy clays and encircled by high ridges with hardly a flat piece of ground. A land well suited to cattle in the lowlands and sheep on the stony upland downs. Not good arable land, but heavily wooded pastureland, just right for nurturing apple trees. It was good cider country where one could live in peace and relative prosperity to enjoy the annual liquid pleasures (as it still is to this day).

The inhospitable nature of the West Dorset coast was also unattractive to potential invasion from the sea. The estuaries of the Axe and the Brit were easy to defend. The waters were too narrow and the countryside too open to risk making a sizeable landing. There were occasional odd exceptions. Sometime between 450 and 550 AD, an important Saxon warrior was buried not far from the sea on Hardown Hill above Chideock. Was this the sad end to a failed invasion? Or did he perhaps succumb to the local cider?

Life in this corner of Dorset continued as it had been for many centuries before the Saxons came. It would have inevitably been very different from medieval life in parts of Somerset, Kent, Sussex, Wiltshire and Hampshire. The residual Celtic influence and its association with the other western European coastal regions was still strong. No doubt there were plenty of comings and goings between coast-dwelling Dorset people and their Breton friends across the Channel; plenty of time to talk about cider, swap a few apple trees perhaps, and put into practice some good ideas gained from hundreds of years of each other's cidermaking.

Thus, excluded to some extent from the rest of England and

from the general evolution of apples in Somerset and other cidermaking counties, Dorset's indigenous apple varieties have remained largely unique. Many have retained a good portion of 'wild' genes, and others have inherited a few characteristics similar to Devon's traditional farm orchard apples. Others, interestingly but not unexpectedly, seem to have acquired a few French qualities. Intriguingly, more of this melange of genetic material is being revealed to us through modern work in sequencing their DNA, often with some exciting results.

The Growth of Dorset Orchards

Dorset's varied landscape, full of contrasting scenery and soils, have influenced social history and farming development, contributing to maintaining its distinctiveness and personality. A big proportion of the county's land is totally unsuitable for fruit culture. In those areas where the soil is suitable, however, orchards once thrived, as described in this snippet from Long Ashton's archives in the 1930s:

Although the acreage of Dorset's orcharding is small compared with that of other counties, it must be remembered that apart from the comparative size of Dorset, the soil in a big proportion of the county is of the type totally unsuitable for fruit culture. In those areas where the soil is suitable however the orchards thrive and are numerously planted and here the production of cider fruit is as much a business as it is in Somerset and Devon.

The largest cider orchard area lies in West Dorset towards the Devon border including and around the neighbourhood of Loders, Powerstock, Netherbury, Beaminster, Broadwindsor and Stoke Abbott. There are smaller areas around Thorncombe, Whitchurch, Wootton Fitzpaine, Chideock and Symondsbury. Most of these orchards are planted in medium loams derived from the middle lias, but there are also quite a number planted in the very light soil of the Bridport sands, in particular around Melplash.

In other parts of Dorset the orchard areas are smaller and widely scattered throughout the county. There is a considerable acreage around Leigh and Chetnole where a good proportion are planted in heavy loams from the Oxford clay. Scattered orchards are found around Gillingham and around Shaftesbury and again around Sturminster Newton, but cider orchards are more numerous in the neighbourhood of Child Okeford, Shillingstone and Hammoon where a good deal of cider is produced.

Yet another area worthy of mention is that around Piddlehinton and Piddletrenthide where numerous orchards are found growing in the alluvium of the narrow valleys.

The enthusiasm for creating cider orchards spread with the popularity of cider. Under Henry II, strong spiced cider was in fashion and especially popular with the monks at Canterbury; two centuries later, in Chaucer's tales, cider is a common and relatively inexpensive drink.

There was a great demand for cider to supply ships going abroad on long voyages, even until more recent days when boats sailed to Newfoundland for the cod fishing. The custom may have begun before the knowledge that cider's vitamin C content was efficacious in combating scurvy, but a daily dose would certainly have helped keep the problem at bay. Selling cider for shipping became a big commercial business, especially in the West Country. Much was sold to ships calling at ports around the coast, some from Bristol but particularly from smaller harbours along the Channel around Devon and Dorset. New Romney in Kent had no less than 13 suppliers of cider destined for shipping. However, many people in England at the time also drank ale. Monastic workers were usually given a daily allowance of cider or beer as part of their wages, a habit that was continued for farm labourers in much of the West Country until the twentieth century.

There was also some two-way sales traffic at times, as we learn from the *Victoria County History of Dorset* (1908). By the reign of Edward IV (1461-1483) cider was being brought into Poole from abroad, possibly from Normandy or even Jersey, where there were large acres of cider apples. 'A vessel batalla named the Mavye of Reyle, Wrench Herbert master, brought in amongst its cargo 1 *pipe* of sidre valued at 3s 4d and Stephen Cressyn, a foreigner, paid thereon 1/2d in customs duty and 2d in subsidy.'

By the early 1500s another vessel, the *Barbaray*, entered Poole Haven under the command of her master Thomas Viron. Amongst the cargo, besides 'great store of apples, pears nuts and other fruits of the earth', were 3 *'puncheons* de perry' containing 1 cask valued at 10s. After disembarking her cargo, she loaded up with English goods and returned home, but reappeared at Poole two months later with more apples and nuts. On this occasion, instead of perry, she brought a hogshead of dry wine, a barrel of verjuice (very likely unripe grape juice for cooking purposes) and two butts of Runnay or Roumey wine which contained '1 cask + 1 hogshead' of dry wine.

England's orchards remained prosperous until the depressions of the fourteenth and fifteenth centuries when, as we learn from William Lambarde writing in his *Perambulation of Kent* (1570), our fruit production was in need of some vigorous stimulation. Happily, it was Henry VIII, always interested in good things, who came to the rescue. He dispatched his fruiterer, Richard Harrys, off to Europe in search of new and better fruit varieties. His successful venture led to the introduction of numerous exciting new apples, pears, cherries and many more kinds of fruit. In came apples with names like French Pippin, Ramboures and Calval. The renown of Henry's model orchard at Teynham in Kent, full of all sorts of healthy, deliciously novel-tasting fruit, was just the incentive required to stimulate the revival of orchard-planting countrywide. Many new orchards were established in Kent, mainly with choice fruit for the tables of the wealthy but also with some cider apples. Though it was in the West Country that serious interest in new and better cider orchards was rekindled.

Between the sixteenth and eighteenth centuries, there was much written about growing fruit, mainly by well educated scientifically minded authors. Those from the cidermaking counties of Herefordshire, Monmouthshire, Gloucestershire, Somerset and Devon were largely wealthy and educated landowners prepared to experiment and voice their opinions on their own and others' successes and failures. The drinking of cider and the planting of cider orchards was also actively encouraged by many influential writers during the times when England was at war with France and good French wine would have been in very short supply.

In his comprehensive volume, *The Whole Art of Husbandry, Or, The Way of Managing and Improving of Land* (1707), John

Vinetum Britannicum:
OR A
TREATISE
OF
CIDER;

And other Wines and Drinks extracted from Fruits Growing in this Kingdom.

With the Method of Propagating all forts of Vinous FRUIT-TREES.

And a DESCRIPTION of the New-Invented I N G E N I O or M I L L, For the more expeditious making of *CIDER*.

And alfo the right way of making METHEGLIN and BIRCH-WINE.

The Second Impreffion, much Enlarged.

To which is added, A Difcourfe teaching the beft way of Improving BEES.

With Copper Plates.

By *J. Worlidge.* Gent.

L O N D O N,
Printed for *Thomas Dring,* over againft the Inner-Temple-gate; and *Thomas Burrel,* at the Golden-ball under St. *Dunftan's* Church in *Fleet-ftreet.* 1678.

Worlidge's *Vinicum Britannicum*, 1671, a text book on cider for his time, and illustrating his proud invention, the Ingenio or fruit mill.

Mortimer, a Fellow of the Royal Society, referred to comments made by earlier horticultural writers and cider devotees, John Evelyn in *Sylva* (1664) and Ralph Austen in his *Treatise of Fruit Tree* (1657). They asserted that growing apples and making cider would not only make our land more profitable by yielding a crop of both apples and grass, but that it would 'hinder the vast consumption of French wines, which is enriching of a foreigner by a trade very prejudicial to this Nation.'

The good John Mortimer then went on to say: 'The use of fruit is also universal both for eating and drinking . . . especially the juice for cider, which being made from good fruit and well prepared is a most delicious, wholesome liquor, and most natural to our English bodies, there being no county in England that hath afforded longer-liv'd people than the Cyder Counties.'

Horticultural knowledge and expertise were growing fast, encouraged by each volume that was published with aphorisms on the ideal means of establishing cider orchards and detailed personal accounts of making cider. For the first time we learn a little about what apple varieties were used, how best to grow them and how they might be blended for cider perfection.

West Country cider orchards

Herefordshire has always considered itself to be the leading cider county, but orchards were also well established in Devon, Somerset and the Channel Isles for just as long, although there are few early records. Celia Feines, Wiltshire-born connoisseur of fruit and cider, travelled extensively on horseback throughout the fruit-growing areas of England in the late 1600s. After visiting the West Country in 1696, she remarked somewhat disparagingly that it was very fruitful of orchards but 'they are not curious in the planting of the best sort of fruit . . . they are likewise careless when they make cider, they press all sorts of apples together, else they may have as good cider as in other parts, even as good as Herefordshire.'

We learn more about trees in Devon from the writings of Hugh Stafford's *Treatise on Cidermaking* (1727), especially about the orchards that covered the South Hams, the area with red soils between the rivers Teign and Dart. Every valley was said to be filled with orchards and the cider excellent. Orchards were also common in parts of central Devon, in the pleasant climate around the Exe valley and from Exmouth to Axminster and into Dorset. In Hugh Stafford's time, West Country cider was much sought after by the London market. It was easy to transport by sea and large quantities travelled by small boats and ships from various ports and estuaries along the Devon and Dorset coast. Places like Axmouth and Salcombe may have seen many liquid cargoes on their way to London and the southeastern counties.

Many small farms welcomed the annual visit of a travelling cider press. All hands help load the apples while the 'boss' turns the scratter mill.

Traditional West Country orchards were usually small, as were the closely planted trees, their size dictated by local soils and topography. The grass was grazed by horses and more often sheep, never by cattle, as the tall, widely spaced trees of Herefordshire's orchards allowed. But what the orchards lacked in size, both in Devon and Dorset, they made up for in numbers and since most farms had their own apple supply, the overall acreage was considerable. Over the years the borders of Dorset have strayed in several places, especially the far western boundary, which at one time encompassed a considerable portion of east Devon. I feel it is safe to say that most references to Devon orchards in the past, if not specifically referring to places like the South Hams, would have included that part of Dorset that was once on the other side of the wandering line of the river Axe.

There are not many early records of farm orchards since they were generally accepted as an essential part of the farmstead design.

However, they were sometimes considered worth recording when properties were changing hands, especially for sales of farms in areas of good productive land. When Befferlands Farm on the river Char, near Charmouth, came up for a sale in 1769, it was described as a 'compact, handy farm – on exceeding good land.' It boasted 35 fields of mostly pasture and two orchards. Doghouse Farm on good land near Chideock still had two orchards in 1897. Although Stonebarrow Farm was mainly down to pasture, it did have four small orchards at one time, one in the farmyard and three others planted in the meads further up the hillside. Most probably their apples supplied cider for the pubs and houses of nearby bustling Charmouth.

One of the most interesting records of cider orchards comes from the Ship Inn. This was for many years a reputable dining house just south-west of Chideock, a renowned stopping-place for travellers on the road from Dorchester to Exeter. It boasted three orchards to supply its customers with good local cider. Sadly, when the route of the main road was altered in the mid 1800s, the Ship Inn became by-passed, its trade dwindled, and its orchards disappeared.

John Claridge wrote in 1793 in his *Agriculture of Dorset*: 'There is a considerable quantity of orchards in the Vale of Blackmore and on the Somerset and Devonshire side of the county, and the cyder made is mostly of the Devonshire sorts. It is chiefly used for home consumption, and I heard of no plantations sufficiently extensive where the grower could sell to other countries to make a considerable return.' The Blackmore Vale was still highly prized for Cider in 1802.

The *Victoria County History of Dorset* (1908) tells us that in 1788, 'Apples were raised in abundance on the land lying between Charmouth and Bridport, the cider from which being sold at 7s to 12s a hogshead.' And from Shaw's *Tour of England* in 1793, 'cider sold for the goodly sum of one guinea to 30 shillings a hogshead of 63 gallons' (larger than the normal 56 gallon hogshead).

We learn so much more from the *General View of the Agriculture of the County of Dorset* published by the Board of Agriculture. In it, William Stevenson reported that in 1812 there were some 10,000 or

Time for a cider break during haymaking. Note the small child holding the cart horse steady.

more acres of orchards. He tells us in detail about some of those that he saw and even gives the names of a few cider apples and people, like Mr Roberts of Burton Bradstock who 'has planted a small orchard on what he considers to be a very improved plan. Large holes, I believe as much as three feet deep, were dug, and half filled with furze and rubbish, previous to the good mould being put in which surrounds the roots of the apple trees. The trees planted in this manner appear to have grown very rapidly, as is the case with the quick hedge which surrounds this garden and orchard, the subsoil of which was also mixed with furze and rubbish, previous to planting of the quicksets.' He laso names Mr Groves of Abbotsbury, who 'thinks it improper to support apple-trees with props, after they have become firmly rooted, and has convincing proofs that they thrive best when the roots are moderately shaken and loosened by the wind.'

William Stevenson tells us just a little about how Dorset cider is made on some farms, and tantalizingly gives a few names to the apples that went to make it. 'In the neighbourhood of Sherborne, it is common to mix six bushels of sweet apples with three of sharp or bitter sorts, in making cyder; and in some places a few crabs are substituted for the rough or bitter apples. In Dalwood the best cider was made from Bitter Jersey, Buckland Marylebone and Langstone apples . . . At Bradford Abbas, and some other places, it is asserted that 20 bushels of apples will make a hogshead of cyder. And Mr Strong of Powerstock observes that an apple tree in that neighbourhood has been known to produce seven hogsheads in a season.'

There was a good quantity of cider made and drunk, especially on farms and particularly at haymaking time when quantities of farmhouse cider were essential. In the early 1800s, Netherbury labourers made 2/6d a day and 12 pints of cider. The Dalwood harvesters got a half pint of cider for every ridge they went down.

Mr John Otton was tenant of Wootton Farm, part of the Manor Farm estate in Wootton Fitzpaine in the early 1800s. The farm boasted a 20-acre nursery on some of the best land in the Vale, with trees planted 20 feet apart. Although the yield varied considerably from year to year, his trees usually made 10 hogsheads of cider (560 gallons). There was a clause in Mr Otton's lease that instructed him to keep his cider mill and press in good order. During harvest, he paid his labourers 7/6d–9s per week. At harvest time the men worked from 6am till midnight and were allowed 2 gallons of cider a day. Women worked shorter hours for 6d a day, but this went up to a shilling a day plus 3 pints of cider during harvest. Wheat was reaped and tied at 7s per acre, other corn was mowed with a scythe for 1/6d per acre and grass was mowed at 2/6d per acre, all plus ale or cider. Providing sufficient cider for the annual farm consumption was an important duty and the process of harvesting the apples, pressing them and fermenting their juice to the required standard to please the farm labourers, took a deal of care.

Dwindling acreages

Once Napoleon had been vanquished and the wars with France had ended, the large-scale trading of French wine returned in earnest, accelerating the demise of cider production and orcharding. This was especially felt in the Midlands as tastes and fashions turned back to wine-drinking and cider was eschewed. In the early 1800s, the recently formed Board of Agriculture made a nationwide survey of farming in a bid to 'improve' our country's efficiency and productivity. Their reports tell us the overall condition of orchards was declining. Herefordshire's cider heyday was passed. Old and once valuable fruit varieties were being lost and most orchards were in dire need of considerable renovation. Even Somerset's orchards were in a better condition than Herefordshire's.

In the early nineteenth century, William Stevenson is credited with saying: 'Of the many counties I have critically inspected, Dorset has proved one of, if not the most disappointing. In all directions small and occasionally large orchards are to be seen, no holding or farm apparently being complete without one, but the trees are truly in a wretched plight, to be matched for their drapery of moss and lichen in other parts of the county, but surely not out of it.' In truth, the 'drapery' he reported was a credit to the clean fresh breezes of the nearby sea.

The Board of Agriculture's estimated acreage of orchards around the country published in 1877, was the first of its kind. It seems that Devon had more orchards than any of the other counties, topping the estimates with 23,000 acres. Herefordshire was second with 22,000 acres, Somerset with 21,000. Worcestershire and Gloucestershire combined had 17,000 acres, and Kent around 6,000, although these three counties mostly grew apples for market. Dorset boasted 10,000 acres of trees.

For its size Dorset was on an equivalent footing to the other counties but its cider output dwindled markedly before the twentieth century. This was partly due to the dominance of local breweries in Blandford, Dorchester, Weymouth, Bridport and Gillingham, and the lack of cidermaking on a commercial scale.

Most orchards were family holdings that produced small quantities of fruit either for themselves or for local pubs and farm-gate sales. Also, in parts of Dorset the commercial prospects of growing flax and hemp were far more profitable and much of Dorset's better land was occupied with growing the raw material for the thriving rope, sail-making and netting industry.

Crab apples for rootstocks

In the early nineteenth century, a Mr Groves who had a nursery in Bettiscombe, sold his apple trees at 3/6d each. His trees were grafted onto crab-stocks nine or ten inches from the ground, with 'clay wrapped round them in the usual way'.

Crab apple seedlings have been used as stocks for apple trees for hundreds of years, even up until the middle of the twentieth century, as a method of raising a desired variety that could not otherwise be grown on its own roots. In addition to providing support, a rootstock will always impart some of its own character on the scion, sometimes by altering the vigour, sometimes by improving the flavour of the fruit. Our modern apple rootstocks are graded in vigour so that the strength and final size of the tree can be controlled: small garden trees will be on the dwarfing M9 or even smaller M27 rootstocks; bush cider trees usually have a MM106 or MM111 stock; standard trees might be on M25, a bold stock with strong root anchorage.

John Mortimer considered that the best rootstocks grown in Herefordshire were those raised from the seeds of the healthiest looking crabs or wildlings. 'Where you can get crab stocks enough in the woods, you may plant your nursery with them.' But for large-scale propagation, stocks were raised from seed and the best chosen.

Crab apples were also often used for cider either alone or mixed with other sorts. As Mortimer wrote a few years earlier, 'Crabs when kept till they are mellow may be reckoned amongst apples, and being ground with other mellow fruit do much enrich the Cyder, and is the best refiner of foul Cyder.' By the time they had mellowed they would have lost much of their harsh tannin. He suggests that crab

The Redstreak known famously as Scudamore's Crab, from Thomas Andrew Knight's *Pomona Herefordiensis*, 1811. This was undoubtedly a seedling crab, but its cider was of such perfection that Knight maintained that every orchard should have a Redstreak.

apples were often grown specifically to be used for cider even if they were 'wild and harsh, not at all tempting to the palate of a thievish neighbour.' A natural and useful anti-scrumping strategy.

Herefordshire already had a list of primitive cider apples in the early eighteenth century: Redstreak, White Must, Green Must, Red Must, Gennet Moyl, Eliot's Stocken Apples, Summer and Winter Fillet, the Broomsbury Crab, Olive Underleaf and Fox Whelps. Mortimer gives us a little taster: 'The greater part of them being merely savage and so harsh that hardly swine will eat them, yet yielding the most plentiful, smart and vinous liquor, comparable if not exceeding the best French wine,' if given three or four years to mellow!

Some of our traditional Dorset apple discoveries measure up to eighteenth-century Herefordshire's fierce potency standards. At Marnhull Mill, the two little trees we found growing together in the mill yard could well be 'ancient named' local cultivars. We have no records of who planted them or what their names might have been, but both varieties have worthwhile qualities, especially Hunter's Ground, which has an individual and complex flavour. These two sorts have all the qualities that those early cider apple writers like Mortimer spoke about as being all the rage. It is to be hoped that, given three or four years to mellow, our finds might also yield 'plentiful, smart and vinous liquor.'

But over the years since Mortimer's time, it seems that cider tastes have changed much. Thankfully, bittersweet cider apples are still with us to please the connoisseurs of true west country cider – some consolation in a time when the larger commercial cidermakers are turning towards sweet, bland and easy-to-drink beverages.

Specialist nurseries

During the nineteenth century there were already several specialist nurseries throughout the country, boasting long lists of fruit trees. Miller and Sweet, nurserymen, seedsmen and florists of Bristol, sent out an extensive catalogue of flowering plants, forest and fruit trees in 1808 with an impressive list of 113 apple cultivars, both old varieties and others only recently introduced. They were clearly able to supply an important part of the market that satisfied the growing popularity of private gardens and orchards. Several apples they listed were well known sorts that Robert Hogg wrote about in his *Fruit Manual* of 1886, such as Stubbard, Court de Wyck, Ribston Pippin and Lucombe's Pine.

This last variety, said to date from around 1800, was raised by William Lucombe, a pioneer of fruit hybridization who founded

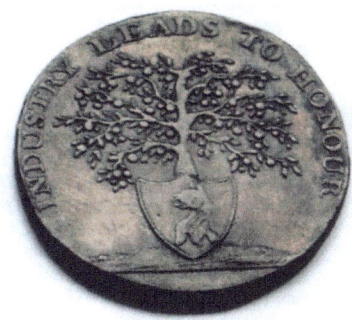

Raising crab apple rootstocks was big business in 1796, as the caption on the Morse Tokens awarded for industrious production demonstrates, 'Several thousands young healthy & fine crab apple & pear stocks raised from the kernel to be sold by J Morse Newent Gloucestershire'. Courtesy of Jim Chapman, Gloucester Orchard Trust

a nursery in the St Thomas district of Exeter in 1720 that became Lucombe, Pince & Co. in the 1820s. Miller and Sweet also supplied several cider apples, varieties now most probably lost to cultivation like the fabled Coccagee, Red Musk, Redstreak, Herefordshire Underleaf, Hagley Crab and White Styre. These are all in Hogg's *The Apple and Pear as Vintage Fruits* with a strong recommendation that they are well worth acquiring to replace 'those older sorts that are now lacking in merit to the detriment of the cider they produce.' Hogg's *Fruit Manual*, stuffed full of recommendations and descriptions, including notes on growing top and soft fruits, such as grapes, nuts and (even more exotic) nectarines, illustrates admirably the breadth of interest and knowledge in late-Victorian times for those who might want to join in the pursuit of fruit-growing.

Like many West Country villages in the nineteenth century, Loders, near Bridport, had several market gardens, some of which may have produced apple trees along with other produce. It is likely that the pleasant flavoured Symes Seedling is one that was born in Loders – it was certainly grown for sale there at one time and some of its progeny are still growing happily in the garden of Loders Court.

Scott's Nursery in Merriott, Somerset, began growing fruit trees in around 1852 when John Scott bought a market garden. With it came extensive fields of good Yeovil sand, the Somerset equivalent of Bridport foxmould (the same excellent growing material that so much of the villages around Bridport and Beaminster, including Loders, were blessed with). In the 1870s, his extensive catalogue, *Scott's Orchardist,* listed an enormous collection of eating, cooking and cider apples selling at 2s.6d. to 3s.6d for dwarf trees and 1s.6d to 3s.6d for standards. But Golden Ball was the only Dorset variety amongst them.

We occasionally found the vintage bittersharp Somerset variety, Cap of Liberty, down in Dorset. The famous bittersharp cider apple, Porter's Perfection, was raised in Porter's own nursery in East Lambrook sometime around 1880, possibly from a Cap of Liberty cross. I believe Porter's Perfection was developed by design to be introduced at a time when cidermakers were searching for varieties that could vie with Kingston Black's vintage qualities. Porter's has all the best ingredients in its juice and, although disconcertingly late-maturing (in November), it is avidly sought by those patient enough to wait for its harvest after a long, warm summer. Many of the cider varieties that were raised or bred at Porter's nursery have had an enormous influence on the fame and longevity of Somerset's cider heritage and its distinctive bittersweet flavour.

Beaminster's historic orchard

By chance I came across a very interesting snippet of information through a lady who had once lived at Farrs, an imposing stone residence near the centre of Beaminster. Farrs was at one time owned and occupied by Richard Symes, a barrister. An inventory was made of the apple trees that were planted in his orchard in December of 1767. They are all very old varieties and probably long extinct, however the record gives some clues as to the sort of apples that might have been growing in the garden of any residence belonging to a well-heeled Dorset family of the time. Judging by their names, the first three are most probably local varieties: 'Broad Faces', suggests a large culinary type of fruit; 'Flood Hatches' might refer to the place where the tree first arose; and 'Crabb Redstreaks', surely a sharp-tasting wildling crab apple very like one of those eighteenth-century Herefordshire cider apples. There were also Devonshire Dusans, very late-keeping apples that reputedly stayed sound for 'deux ans'. The multipurpose English Pearmains and Pomeroys were popular well-flavoured late season apples.

There were also two very choice vintage cider apples in the Farrs orchard inventory, Golden Pippin and Royal Wildling. The Golden Pippin, a small yellow apple, extensively planted before the end of the seventeenth century, was described by Thomas Andrew Knight in his *Pomona Herefordiensis* as 'a very prime cider apple'. The Royal Wildling, was the legendary variety found as a seedling growing near Exeter, in the parish of Upton Pyne, and brought to fame by Hugh Stafford of Pynes House who extolled its virtues in his *Treatise on Cyder-making* (1727). Royal Wildling juice was described as having a 'delicate roughness and a fine vinous flavour'. He records that 'The gentlemen of our county (Devon) are now busy almost everywhere in promoting it, (as are) some of the wiser farmers. I have known five guineas refused for one of our hogsheads of it, though the common cyder sells for twenty shillings, and the South Ham for twenty-five to thirty.' Later, Robert Hogg added that the Royal Wildling juice was, 'deficient in flavour by itself but its value is from the body and strength it gives to the cider when mixed with other varieties whose juices supply a higher flavour.' Best for blending! It achieved widespread fame and popularity from cider aficionados in the late eighteenth century.

Undoubtedly, this was a collection of apple varieties that might have been chosen by those who could afford the best sorts for eating and drinking at that time. Might Richard Symes have obtained his Royal Wildlings and Golden Pippin's from Hugh Stafford of Pynes? Like many wealthy, educated people at the time, he may well have had colleagues, family and well-to-do acquaintances around the cider counties with their own orchards who passed on scions of heritage trees and exceptional varieties.

Collecting apples, Powerstock, Dorset, *c.*1989, by James Ravilious.

Bound to the Soil

Much changed in the Dorset countryside during the centuries following the Norman invasion: the threat of Black Death, the dissolution of the monasteries of Forde, Cerne and Abbotsbury, the loss of Commoners rights to gather wood and graze their animals. Slowly, broad acreages of farmland on Marshwood Vale's best soils were swallowed up by succeeding generations of wealthy, landowning families that became even more prosperous while there remained a steady price for wool and a good demand for grain. Families like the Coxes of Whitchurch owned a considerable acreage and many separate farmsteads up until the end of the eighteenth century, and made much money from flax and hemp. But life was hard for the small farmer on the poorer land and even harder for the landless farm labourer.

Flax and hemp have been grown and worked in West Dorset for rope-, sail- and net-making since King John ordered hempen thread for ship's cables in 1211, but the industry had its heyday during the Napoleonic Wars when a bounty payment was introduced to encourage farmers. There were probably around 4,000 people directly involved in the industry at its peak, either growing the crops, processing the raw materials or making the ropes, sailcloth and nets. By 1796, there were 102 acres of flax and hemp growing in Chideock alone and 54 acres in Whitchurch. The prosperity generated by this lucrative industry most unquestionably benefitted cider consumption and gave time, perhaps, to relax together at the end of the day with a mug of cider, appreciate its taste and discuss the desirable qualities of this year's product and be grateful for the different flavours of cider apples and apple trees.

By contrast, the 1844 tithe census gives a clear illustration of the difficulties of scratching a living from small fields on arduous terrain. In the broad parish of Whitchurch, most of the holdings

Marshwood Vale's patchwork of cottages and tiny fields.

were small (many less than 10 acres) and only one-third of the parishoners actually owned their land. Only a score of farms across the Vale held more than 100 acres, such as Cards Mill and Abbotts Wotton which also occupied the better land. A farm of that size would have employed many workers and an orchard would have been an essential and well-cared-for possession that had to supply the all-important daily cider allowance to supplemented the wages.

Much of the less productive land was offered for rent to those who could afford it. Sometimes, a piece came complete with cottage, dairy and cows for an annual rent. Others may have had a few acres capable of supporting small herds of just 20 or so cows. This insecure, transient way of living could change swiftly from year to year.

Even during the times when the poor were getting poorer and the rich getting richer, there was a significant number of 'middling sort' of farmers tilling between 10–15 acres in Marshwood Vale. They were able to survive at a tolerable level but may not have owned their land and life would have been a struggle. Some augmented their income as tradesmen and jobbers, perhaps carriers or factors of farm-produced butter. Some may have had a market garden. Undoubtedly, many made ends meet through employment in the sail, rope and net industry. But most did not have the luxury of growing apple trees or developing an orchard. In times of relative affluence, they may have considered planting a tree or two in the corner of the farmyard that would be enough for a little cider to make life brighter, or perhaps a handful of trees to shelter the young lambs in the spring. Their trees could have been grown from seedlings that germinated from the pomace after cidermaking, or perhaps a cutting from a neighbour's tree. They would have been nothing special and certainly not have been purchased. Through a little enthusiasm and encouragement, those apples that tasted good and made the best cider became locally popular 'varieties' – the continuing happy story of apple evolution in action. By serendipity and selection over the years, new varieties of apples emerged, perhaps some of those Dorset ones keeping a few 'French' genes.

Twenty-first century orchards

One October morning, I went to Crabbs Bluntshay Farm on Mutton Street, near Shave Cross, to talk to Malcolm and Sylvia Creed (author of *Dorset's Western Vale*). They have made farmhouse cider for many years, but sadly ill health has put paid to their activities. It was good to see the orchards on the farm, still growing strong and cropping apples every year. It is such a pity that there's no more cidermaking, but it was good to have a look at the cider barn where it all happened.

There was an old, refurbished press and motor for a scratter mill. Without that it would have taken five men to turn the massive handle! The air still smells good in there. Remembered aromas of Kingston Blacks, Pounds, Chisel Jerseys, Morgan Sweets, Dabinetts, Crimson Kings and Golden Balls that had once filled the space. There were still the old dairy equipment and milk coolers in the next barn, both a museum of antique machinery.

Sylvia took us out into the orchards along the side of the river

The old cider press at Stubbs Farm, Marshwood Vale.

Char. There was an old tree in the first orchard that had fallen over a while ago. It was lying in the grass, regenerated, a true 'pitcher' with burr knots all along the length of its trunk rooting into the ground. It was a Kings Favourite, a very Dorset variety. Most of the trees in the other orchards, mixed with the cider sorts, were eaters and cookers, modern varieties like Fiesta, Ellison's Orange and Red Worcester that were planted for the farm shop and campsite.

The cider apple trees were clearly very old, rough barked, many of them leaning into each other for support – a few bittersweet jerseys and a sweet Woodbine, struggling with a crab rootstock that was beginning to take over its crown. The Tom Putt was easy to recognise, as was the Bulmer's Norman, but I couldn't put a name to many of the others. We did take a bite to taste most of them. Some had a good flavour, usually quite sharp. But there was one very interesting old tree, full of green fruit, growing twined together with a holly tree – a bittersweet cider apple of unknown origin. It would have been a good one to blend with some of the sharper apples for cider.

As Sylvia remembered, farmhouse cider orcharding was always a low-cost operation. Very little pruning was ever done, the trees were left to their own devices. Every now and then the brambles would have to be cleared. If a tree fell down, it might be used for firewood or it might be replaced with a young one. Although, in the past, if the farm was rented the tenant was expected to keep up the orchard as a viable concern. She told me that a few of her younger trees had been bought years ago but their names had been forgotten. Some others had come from cuttings given by friends and neighbours as 'good trees for cider', but they rarely had a name. Swapping graftwood is such a familiar custom, performed since time immemorial between enthusiastic gardeners, orchardists and cidermakers alike. I am certain that is how most of the apples in Marshwood Vale were distributed. Their origins may have been lost in time. Perhaps they started life as a seedling in the hedge, or perhaps germinated from the pomace heaps beside the cider barns and presses.

There must have been so much apple wood that changed hands in that manner when the Romans introduced their ancient varieties long ago, and again when the Breton and Norman immigrants brought their own favourite cider apples with them to England. All this exchange over the centuries has given many opportunities to add novel genes to the apple world. Surprisingly, many of our well known varieties of eating, cooking and cider apples have come from seedlings. Certainly, more have begun life in that way than from cross pollination between selected cultivars.

Just a few minutes from Crabbs Bluntshay Farm, hidden in the valley on the other side of Bluntshay Lane, are two farmsteads with orchards running down to either side of the river Char. In one of those orchards, we found Sweet Alford and Tom Legg trees. Their fruit goes most years to Russell Crocker and his brothers to make cider on their farm in Kingcombe. The orchard on the south side has had some generous replanting in recent years. This farm is the home of the famous 'Will's Surgery' that featured in the hilarious film made at the Dare family's farm in the 1960s by Clive Gunnell for Westward Television.

By the end of the nineteenth century, many of the farmers in the Vale had their own cider mill and press. Smaller farms without the gear may have taken their fruit to their neighbours for pressing or waited for the travelling press to come their way.

On the day I visited Char Valley Farm, at Owl Corner in Blackney near Shave Cross, the orchard was home to twelve rams curious to know why I was interested in their field. Only three trees were left of the once extensive orchard: a very old Bramley and two stunted, nibbled trees with striped red and green apples, dual purpose but with a dash of bite in the flesh. Sadly, they are gone now, both rams and trees.

There are many more little orchards like that down the lanes around Mutton Street. Fortunately, there is a lot of recent interest and enthusiasm for getting these orchards back into health and apple production. They are after all a valuable historical record of unique local apples and traditional cider flavours. Some orchards have already benefited from a bit of pruning and renovation, along with some expert advice for their future wellbeing from Nick Gray, West Dorset Conservation Officer of Dorset Wildlife Trust. Most need a few replacement trees. For others, with only a few sheep-gnawed remnants, or perhaps just a lone tree, it may be too late for help. Sadly, often the only trees that are left are just cooking or dual-purpose apples. True cider apples are a rare find.

French influences on the Good Land

From a Long Ashton Report of Dorset cidermaking, written in the late 1930s:

> According to many local farmers, Dorset was the first county in England to make cider. It is claimed that the art of cidermaking was first introduced into this country by monks from Northern France

who settled in a village near Bridport some time before the Norman Conquest. Whether this be true or not, and although there are few records available which deal with the history of cidermaking in the country, Dorset certainly ranks with the counties in the West of England which have produced cider for centuries.

The first 'commercial' apple orchards were planted by the Norman monks after the Conquest. Many monasteries in England had their own *pomeriums*, their fruit orchards. The plan of the monastery in Canterbury in 1165 show the apples trees, many of which would have been carried to England by the monks from the Continent. In 1570, William Lambarde writing in his *Perambulation of Kent*, recorded how the many varieties of fruit had been introduced to England: 'Those plantes which our ancestors had brought hither out of Normandie.' Although they were still in existence three hundred or more years later, he felt that they had now lost some of their 'native verdour'.

Legend has it that it was these very monks of Loders Priory near Bridport who first brought the art of cidermaking to England. They certainly had an orchard in the fields below the Priory. It was on one of the rare flat pieces of land in the valley, not far from today's church of St Mary Magdalen. You can see where it was by looking south over the churchyard wall to where the railway used to run between Powerstock and Bridport. The whole of Loders village is on good land, the best Bridport foxmould. The sandy soil in the deep banks of the footpath up Boarsbarrow hill beyond is soft and truly fox-coloured in places.

There has been a church in Loders since Saxon times. The original was probably constructed over a number of years within the grounds of the old Manor House, where Loders Court is now. It would have been considered the property of the local lord of the manor. The modern church was begun on a larger scale in Norman times in its present position. A little of the original Saxon building remains there today – arch of two great stones leaning together above a walled-up doorway.

The village of Loders has several mentions in Domesday. One

Detail of the Victorian stained glass window in Holy Trinity Church, Bradpole.

records that '*Lodres*' comprised 28 villagers, 24 smallholdings and 9 slaves, and that in 1086 the taxable value to its lord King William was £34, a considerable sum that indicated its prosperity. Most of these holdings would have followed the course of the river Asker, as the houses and buildings do today, a line of small self-sufficient farmsteads spread along a fertile valley. Like Whitchurch Canonicorum, Loders was an unusually large 'parish' for those days, but there the similarity ends for Loders is blessed with good soil, deep, golden foxmould.

After the Norman Conquest and many years of unrest later, King Henry I rewarded one of his stalwart knights Richard de

Redvers for his loyalty with large grants of land in Devon, Dorset and the Isle of Wight. These estates included the manor of Loders with Bothenhampton. Richard in turn gave the manor to the Benedictine abbey of St Mary de Montebourg in Normandy, an abbey that held many significant estates throughout England, including locally a chapel in Bradpole, another priory in Axmouth and possibly Powerstock Church. Loders then became a priory when Richard died in 1107. The monks were Norman French-speaking 'Black Monks', or Benedictines, from an order renowned for progressive farming and with the necessary horticultural skills to make and sell their produce, including cider.

An account of the rents, services and customs of the priory of Loders and Bothenhampton dated the year 1305, notes that several of the tenants were bound to 'gather the apples for making cider'. Whilst working for the priory they were allowed a certain number of bottles of cider or, if they preferred, beer. It is interesting to note that fermented drinks were '*bottled*' at that early date. However, since the reference comes from an old French text, '*bottle*' would have been a direct translation of the French '*boutaille*', a word which could refer to either a glass or earthenware vessel for carrying liquid. However, it is more than likely that these '*bottles*' were made of leather for taking out into the fields, or earthenware open-necked jars that could be stoppered. Although there is no mention of where the mill at Bothenhampton was located, there is a note that suggests the watermill might have been used to crush the apples. The monks themselves very possibly drank wine rather than cider, at least some of the time. There is a record of a barrel of wine that arrived from France by sea to Loders via Bothenhampton in 1261.

It is certain that the first immigrants who came over to live and work with the monks on their estates brought the best fruit varieties with them from the Continent, including their own cider apples – ones that made cider to their own home-grown taste to add to those that were already used here. And being familiar with the requirements for making good cider, there would have been significant planting of apples solely for that purpose.

Among the English apples that were grown in the monasteries and manors of thirteenth-century England, the Old English Pearmain was the first to be recorded, in a deed of 1204 in Norfolk. It was an apple prized both for dessert and for making good quality cider. Lord Runham's manor was required to pay the Exchequer each feast day of St Michael, 200 Pearmains and four hogsheads of cider made from them for the King's pleasure. The Costard (also a character from Shakespeare's *Love's Labours Lost*) is another English apple, a cooking apple also well known in France, which was recorded as being sold in Oxford in 1296 at a shilling for 100 fruits. It was popular for making pies until after Shakespeare's time.

Although the monks in many monasteries made cider for their own use and for the workers on their estates, the majority was intended to be sold outside the monastery, as they did with their other farm produce, and was documented in their accounts. Battle Abbey in Sussex recorded theirs in 1275 and in 1352, when the almoner at Winchester was required to report the failure of that year's apple crop. A few years later, Battle Abbey sold three tuns of cider, large casks that may have held over 250 gallons, for 55 shillings each. The first reference to cider in Dorset is from a Pipe roll of Edward I in 1291, which mentions *cisera* in an account of Shaftesbury Abbey. The Nonarum Inquisitiones of 1340 includes tithes of cider in the parish of Beaminster. Sadly, no records of the apple crops or cider sales at Loders have ever been found; if any records were made, they may have been lost when the mother abbey of St Mary de Montebourg in Normandy was destroyed in the Second World War.

In those early centuries following the Norman Conquest, manors and monasteries enjoyed considerable prosperity from the sale of their orchard produce, perhaps more for the cider they yielded than for the apples. The Priory of Loders, like many other houses, was seized by the King several times during the period of the French wars. When Loders was seized by King John in 1204, the land was reported to be worth £33; £40 with all its stock. This goodly sum suggests a level of prosperity. The following year the sheriff was ordered to restore to Prior Baldwin full possession of his property, for which he had given two palfreys to the King, and a promise to pay to the King whatever he had formerly paid to the Abbot in France, and not to transport any goods abroad without a licence.

The Benedictine abbey of St Mary de Montebourg held the priory church until 1414 when it was suppressed, and the land handed on to a nunnery near Twickenham until the dissolution in Henry VIII's time. By then, Loders was a sizeable possession with rents amounting to over £120. Although there is no sign of the original priory now, there are some remains of an ancient stone building near the church where historians believe it may have been.

It is interesting to ponder the local legacy that Loders priory could have left behind. There are a few French names that occur in the local parish registers, such as Malachi Daubeny, the tenant of Marlpitts Farm, West Milton, in 1871. This surname is of Old French origin, a variant of the ancient and distinguished Daubeney, and a locational name. The original Daubenys could have come from a number of places in northern France, including Aubigne in Brittany or Aubigny in Normandy. One of William the Conqueror's attendants was a Daubeny. He came from Manche in Normandy, and he was one of the first to have his family name recorded.

For the 300 years or more that monks lived and worked at Loders, there were French apple varieties growing in the priory orchard and most probably also in the village of Powerstock. Undeniably, their progeny spread and prospered in farms and gardens throughout the surrounding villages. Were some of the trees that we discovered their modern descendants? Their French DNA certainly suggests that they could be.

A portrait of Loders

Thanks to the help of kind members of Loders History Society who unearthed details from their Tithe Apportionment and parish records, we have an invaluable insight into the life of the people of Loders, a pretty much self-sufficient village in 1846. We have a snapshot of life in some of the farm and cottage holdings; we learn about the villagers' occupations, their crops and begin to appreciate the significance of Loders' inherited soil fertility. We also get a clear view of the abundance of orchards in the surrounding countryside.

How good the apple harvest must have been! How splendid the cider must have tasted!

The land either side of the River Asker and up on the surrounding hillsides is excellent for mixed farming. In 1845, the parish of Loders supported 1,082 acres of arable land, 986 acres of meadow or pastureland and over 100 orchards in and around the village. Sir Molyneux Hyde Nepean, Bart., who lived with his family at Loders Court, owned a considerable number of properties and over half of the land. Many of the cottages on the south side of Loders Street were his possessions, as were extensive parts of land on the rounded foxmould hills of Waddon, Symes and Boars Barrow.

After the Apportionment of 1846, occupiers paid rent instead of tithes to the owners of their land. This involved numerous parish meetings and considerable discussion about how fair sums should be assessed, and to whom they would be due. In Loders, the most complicated area of calculation was for those considerable areas of land under arable cultivation. After several years of debate, it was declared that those farmers working arable land in the western end of Loders would pay their annual rent to Sir Molyneux Nepean, and those in the Yondover district to Ann Brown. The actual sum of rent to be charged was calculated on the value per bushel of wheat, barley and oats.

Job Hansford, aged 40 in 1846, and his 35-year-old wife Elizabeth, were working the land at Newhouse Farm on the corner of Barr Lane and Yondover. Their farm was originally called Shoothouse after the water shoot that comes out of the boundary wall on the corner of the road. Job farmed around 38 acres as arable, with 40 more acres as meadow and pasture. He was one of the largest owners of fruit trees in the village, keeping around four acres of orchards in the fields surrounding their farmhouse. He managed another two good orchards on the other side of the Yondover road, squeezed in between those of his neighbours. And

Loders orchards and gardens.

from them would have to pay an annual rent to Ann Brown. There was an additional rent charge for gardens, meadows and pasture according to the number of sheep and dairy cows kept there each year. Consequently, Job's annual rent bills were assessed at £17/10/6, not forgetting a contribution of £9.16 to the vicar for the parochial needs. Job was then free to farm what crops he liked.

Henry Gerrard of Wadden Farm worked a large acreage at the other end of Loders. A good deal of Henry's 166 acres was down to arable, which included at least one three-acre hemp lawn, but there was also over an acre of south-facing orchard at Westcombe and more still higher up towards Symes Hill. There were plenty of fruit trees by his house and garden at the church end of town, plus a further two acres at Home Farmhouse orchard and another small orchard behind a house belonging to him.

Lydia Knight kept a good acre of trees at the village end of Barr Lane. Further on along Yondover, where the land ran gently down to the River Asker, was a row of orchards belonging to neighbours of Job Hansford's: John Hopkins, Charles Salisbury, Thomas Gale (inn keeper at the Farmers Arms who kept an acre of trees) and Thomas Marsh. The Blue Ball inn had its own small orchard and the Three Cups kept an orchard of two rods (60 square yards). The Reverend F. MacCarthy owned a good acre of orchard behind the Loders Arms.

There were never any records made of the sorts of apple growing in the larger orchards, but it is safe to believe that they grew true cider apples destined for milling at the end of the season. There were no cider mills mentioned in the mid-nineteenth-century parish records, but like most West Country villages, the barns of all the larger farms would have stored apple mills and cider presses as they had done for 2-300 years or more. There would be a stone mill, either under cover or in the yard where there was space for a horse to turn it. And inside, a typical oak beam screw press. On outlying farms, or where there was no local cider mill, it was common practice for apples to be harvested and stored in the orchard in sacks ready to be taken to a farm for pressing. Or there was the travelling cider press. The arrival of the traction engine press and crew in the autumn created great excitement, an event that would have brought the younger villagers out to watch the steaming machine go from orchard to orchard milling and pressing as it went.

The Ordnance Survey maps published between 1842 and 1952 show just how many trees there were in the gardens behind the cottages. Although reasonably large by modern standards, they were not big enough for many trees. John Gerard, George Samways, Mary Lathery and Edith Marsh, all had gardens with a few fruit trees, probably mostly apples for family cooking and eating and perhaps a few plums and pear trees. Most cottagers would have kept a pig, a few geese, hens and ducks in the orchard but some open ground would have been needed for a vegetable garden to support a large growing family. It is unlikely that the apples the cottagers grew were destined specifically for cider, but in years of a good autumn crop, plenty would have contributed to the pile at the mill in exchange for a drop of cider in the spring.

The Celtic saint of cider?

A Saxon church was built in Whitchurch Canonicorum towards the end of King Alfred's time. As at Loders, this was rebuilt on a larger scale by the Normans using many of the Saxon stones (and some recycled Roman bricks). Several of the stone carvings on its exterior walls are thought to date from Saxon times: a ship, an axe and a long handled reaping hook. Interestingly, there is a carved drinking vessel said to represent the Holy Grail set into one wall. It is a two-handled vessel that is usually described as a typical twelfth- or thirteenth-century chalice. To a cider drinker, however, it very much resembles a cider mug, which is what it is: an owl jar. These vessels were for labourers to take into the fields – a rounded earthenware, lead-glazed mug, to be worn around the neck on a thong that passed through the jar's two ear loops. (There is one on display in Sherborne Museum.) Perhaps the stonemason who wrought the Holy Grail at Whitchurch Canonicorum was inspired by a common cider mug for his design.

Whitchurch village church hides another secret from a much earlier time: a leaden casket believed to contain the bones of St Wite. A shrine with its saintly relics was once a place of pilgrimage for healing the sick, especially those with eye problems. There are many versions of St Wite's life history, and why reasons her bones ended up in this spot in Dorset. The most likely explanation was pieced together by the Rev. S Baring-Gould, a recognised authority on such Celtic mysteries. Although there is no English legend for her, St Wite is recognised in Brittany under the Celtic name of Gwen, the Latin name of Candida and the French name of Blanche – hence her English name – Wite. She was the daughter of Budic II, a Breton prince (known in Welsh as Emyr Llydaw). She had a son, St Cadfan of Bardsey Island fame, who later founded the Tywyn church in Merionethshire. On the death of her first husband in the late fifth century, Wite married Fragan, a cousin of Cado, Duke of Cornwall. By marrying twice, she earned the title 'Gwen of the three breasts', an expression used among the Welsh and Irish for a woman twice wed and with children from both marriages. The couple soon left

Carving of St Wite at Whitchurch Canonicorum.

Britain and crossed the channel to Brittany, where they settled in Ploufragan (near Saint-Brieuc), the tribal residence of Fragan.

There are several lurid tales about St Wite that add interest and confusion to her legend: she was carried off by pirates to London; that one of the pirates cut off her left hand with an axe when she was making her escape; that miraculously she then walked back over the sea to Brittany, her pathway leaving a visible line across the water. How much her incredible survival of these ordeals contributed to her reputation as a saintly healer cannot be known, but she remains a popular saint in Brittany where to this day the fishermen refer to a line of foam left by the turning tide as 'St Blanche's Causeway'. Some people in West Dorset still call it 'The trail of St Wite.'

There is a possible explanation for her relics ending up here in the

A carving on the church wall at Whitchurch Canonicorum: is this the Holy Grail or an owl cider jar, comparable to the example in Sherborne Museum (*below*).

village church at Whitchurch. Charlemagne, the great Emperor had died in 814 and Europe was once more in turmoil. In the absence of his control, the Vikings invaded Brittany in 914, occupying it for many years and causing great distress and disaster for the people. Many Bretons fled their homeland for England together with numbers of expatriate Britons. They brought their valuables with them, including the relics of ancient Welsh and Breton saints. It is said that Athelstan, although not yet king in England, received any refugees forced to flee from Vikings kindly and allowed them to remain in west Wessex. The bones of St Wite may well have come to Whitchurch at this time. It is said that the carved emblems on the church stonework make reference to her legend: the axe that cut off her fingers, the ship that returned her to England, the hook that represents the pirates.

Hers is a life story as interwoven with myths and legends as a piece of Celtic knotwork. Her memory will not be forgotten as long as the county flag flies, for the red bordered white cross on the yellow ground is St Wite's own.

Although there is much speculation, controversy and conjecture about her legend, there is enough truth to support the likelihood of Breton influence over Dorset's early history. Who knows how much cidermaking and drinking went on in those dark mediaeval days? But what is certain, the apples were here and ready for pressing each autumn, and plenty of eager souls were at hand with the knowledge and expertise needed to make it.

Science for Cider

Cidermaker Edward Smith of Stoke Abbott won a prestigious 2nd prize at the International Brewers Exhibition, London in 1932.

The reputation of cider was probably at its all-time low by the end of the nineteenth century. The Member of Parliament for Hereford, C. W. Radcliffe Cooke, commented that 'cider and perries are all alike and only to be told from vinegar by a highly discriminating palate.' It caused much dismay and serious discussions on how its status might be resurrected. Fortunately, several influential people stepped in to revive cider's flagging reputation, the most prominent and influential among them being Robert Neville Grenville of Butleigh Court in Somerset. He had been working hard for a number of years with his assistant cidermakers, William McCreath and Frank Lloyd, a microbiologist, making and experimenting with cider in his own laboratory. Their results were so encouraging that with the help of the Bath and West of England Society, the Ministry of Agriculture was persuaded to found the National Fruit and Cider Institute in 1903, later renamed Long Ashton Research Station. Its prime objective was to explore the best methods of ensuring reliable produce for all the cidermaking counties.

The Bible of vintage cider apples

Rapid progress at Long Ashton was made following the appointment of Professor B. T. P. 'Berty' Barker, an Oxford don who had made an exceptional study of yeasts. The work that followed under his supervision as Director led to many advances in the understanding of fermentation and the importance of hygiene for cidermaking. From the start, farmers were encouraged to send fruit to the Station's Cider House for assessment every year. Like Neville Grenville, Professor Barker recognised the vital importance of apple varieties and their juice on the flavour and quality of cider. Each sample of apples was pressed, and its juice

Casual cider judging for the Bath and West Show, 1896.

analysed for acidity and tannin content, then classified according to the categories of taste we use today: bittersweet, sweet, sharp and bittersharp. Professor Barker published a list of the most promising varieties in the Ministry of Agriculture Bulletin 104, *Vintage Cider Apples*, which became 'the Bible'.

It was clear that the vintage quality of an apple's juice could not be determined by its chemical composition alone. Only by making and tasting its cider could its true value be appreciated. Farmers were invited to return to Long Ashton to taste their single variety ciders the following year. The first tasting day, in 1906, was attended by only a small group of members, but the annual event later attracted so much interest from participating farmers that by 1924 it was decided to develop the proceedings into a cider competition, when the number of attendees ran into hundreds. Many Dorset farmers who made a little cider must have gone up to Long Ashton for the legendary Annual Cider Tasting and Prize-giving, to taste, to hear the results, and perhaps come home with a prize. It must have been an exciting day out for farming families, dressed in their best, gathering under a massive marquee erected on the lawns with plenty of bread and cheese to go round.

In those early days, most of the competition entries of fruit came from Somerset, Herefordshire, Devon and occasionally from Cornwall. Few entered from Dorset. However, in 1922 Mr H. R. Spence sent in Redstreak and String Pippin apples from his orchard in Stoke Abbott. Then, in 1929, his neighbour J. Bowditch got a commendation for his Golden Ball on its first appearance in the Special Vintage class, as 'an acceptable medium sharp cider'. Praise indeed! There were six Dorset entries to the cider house that year, all from the Bridport area. Bowditch's neighbour, E. C. Smith, sent in Ironsides (French Crab), bittersharp Middlemead and a mixture of Best Bearer and Kings Favourite. Mr Bowditch also sent in a mix of Coccagee and Golden Ball. That would have made a robust cider! We have no idea what Coccagee, the 'goose turd' apple might have added in the way of character, although we can hazard a guess as to what it looked like! It was reputed to make a very 'strong, harsh but excellent' cider and was certainly widely grown in parts of Devon in the nineteenth century, and respected for its individuality. It is still sought out, mainly for its name. Sadly, we seem to be too late to find any trees still surviving but it may still exist in the corner of a field somewhere, who knows?

The Cidermaking Instruction Scheme

A Long Ashton Farm Orchard Survey in 1918 highlighted the shortcomings of the cider business countrywide. Most farmers lacked any knowledge of fruit varieties or how the raw juice should taste. Apples were gathered and processed with no regard as to whether they were ripe and ready to be harvested, or under-ripe, over-ripe or rotten. As there didn't seem to be any knowledge of how to control the fermentation, good ciders were rare. Most could be described as 'haymaking' ciders, suitable for farmworkers hardened to a taste with character. Modern cidermaking methods were only just being introduced into Dorset, and there was little knowledge of individual apple varieties. It was becoming clear that some formal instruction would be necessary.

In 1926, the Cidermaking Instruction Scheme was begun with the object of assisting and encouraging farmers to produce a more palatable type of cider as a profitable branch of the farming industry. Many enquiries came into Long Ashton, wanting to know of the best sorts of apples, and the interest in 'vintage' cider varieties had increased greatly. All these activities clearly demanded the appointment of a proper Instructor, and Phillips Thornley Hyde Pickford took up the challenge. At the start, he compiled a pencil-written list of eleven Dorset apple varieties that were known to be still grown, with very brief notes on the appearance of the fruit: Bell, Best Bearer, Buttery d'Or, Golden Ball (Neverblight), Ironsides, Crimson King (Kings Favourite), Long Stem, Pound Apple, Somerset Woodbine (Runaway), Somerset Crab (Grab) and Tom Legg. Based at Long Ashton, Mr Pickford liaised with county representatives to cooperate with the scheme. J. E. Forshaw, the County Horticultural Advisor, led the work in Somerset while the counties of Dorset, Worcestershire and Monmouthshire found suitable local aides.

There followed a series of extremely popular and successful farm cidermaking courses in all the cidermaking counties. Dorset was host to an autumn cidermaking demonstration, and in the following summer there were 48 farm visits to Dorset alone. In the first year of the scheme, the instructor gave complete guidance for the whole cidermaking process. He selected the apples to an appropriate blend and supervised the fermentation, filtering and bottling through frequent visits to the participating farms.

The annual Long Ashton Report for 1926/7 describes the Scheme as an instant success:

> The services of the Instructor have been in great demand and the value of his work may be seen in the striking improvement in the quality of the cider made by the farmers who have sought his assistance. Some have been successful in obtaining awards at open Competitions for cider and perry at various shows, and it is believed that for the first time entries from Dorset, one of the counties included in the scheme, have found places in the awards list.

Long Ashton's cidermaker, Bob Hathaway (with cigarette) judging the quality of farm cider apple, 1930s.

And that winner was Harry Warren from Netherbury with his Golden Ball cider.

The demonstration lectures in Dorset farms continued over the rest of the decade, with keen interest in making further improvements. Most of Dorset's cider was of a 'low acid type' which always risks disorders like *ropiness and sickness* during fermentation. The instructors spent a great deal of time on carefully selecting the correct varieties and making judicious blends. How much careful selection went on after the instructor had left, is debatable. But one instructor noted, 'although at the start it was found that the art of cidermaking was on the whole more advanced in Hereford and Worcester than in Dorset and Monmouth, satisfactory progress was soon made in the two latter counties.'

Already by the 1930s, cider's reputation had changed remarkably thanks to Long Ashton's experiments and guidance. The main objective had been achieved: to improve the standard of cider that was commonly made on farms at the time and generate a more marketable product. There was a growing demand for a high quality cider from the general public who up until then had not recognised that cider could be a palatable drink! The preferred taste was a more refined sweet or medium sweet cider with a slightly 'brisk' flavour, free from any taint or acetification that characterised many of the completely dry 'farmhouse' ciders. This type of cider had a much higher market value than the rough farmhouse product. Now, instead of being a drink mainly confined to the West of England and largely only consumed in country districts and on farms in the villages, cider had 'come of age' and become a popular drink throughout the British Isles.

The Dorset Federation of Farm Cidermakers

A limited number of farmers were keen to keep abreast with the modern methods of cidermaking, and began to specialise in a drink with a more popular taste. Factory-made cider held the greatest share of the market and many business-minded farmers no longer made cider themselves, but sold off their apples to them. Those who persisted in making farm cider were in a difficult position. The old practice of 'paying' farm labourers with cider had practically died out. In addition, many public houses that would once have taken their surplus cider were, especially in Dorset, tied by the breweries to the cider they supplied.

During this inter-war, the Ministry of Agriculture created several trademarks to assure both the quality of the food concerned and that its origins were British. The National Mark Scheme proved to be an enormous trade advantage to some larger cider producers, but qualification was limited to those whose sales were not less than 36 pints (or 72 half-pints), which handicapped smaller farm producers.

In 1929, the Dorset Federation of Farm Cidermakers was founded as a local branch of the Farmers' Union. (An office was later set up at 21 West Street, Bridport. R M Gillington was the secretary.) The Federation's main objective was to encourage a high standard of local produce, which proved invaluable for small-scale members who did not make enough cider to qualify for the National Mark individually. They were now able to do so under the Federation's umbrella.

Government approval. A poster celebrating and promoting British cider in the 1930s.

THE AUGUST CIDER CUP

1 bottle National Mark cider
2 – 3 slices of National Mark apple
Juice of 2 lemons
1 pint soda water
1 dessertspoonful sugar
Dissolve the sugar in the lemon juice. Put all the ingredients into one large jug. Chill before serving. If a refrigerator is available, place it in the cabinet for 2 hours before serving or add one or two ice cubes to the beverage.
Sliced apple or one or two strawberries or raspberries, should be placed in the cup, as they add to its attractiveness and also improve the flavour.

Extract from *The National Mark Calendar of Cooking* compiled and published for the Ministry of Agriculture in 1936.

Linden Lea Cider Company

Visitors to Netherbury, near Bridport, 20 years ago might remember Hubert Warren, the last remaining true Dorset cidermaker of the old league. His barn in the orchard hid barrels that once held his father Harry's award-winning cider under the name of Linden Lea Cider Company, in acknowledgment of the famous poem by William Barnes 'My Orcha'd in Linden Lea' (1859) was in native Dorset dialect. That little area of West Dorset just south of Beaminster was clearly a hotbed of cider enthusiasts in the early 1900s, all keen to achieve great competition results with excellent produce at the big annual shows – the Bath and West and Southern Counties Society, the Royal Counties Agricultural Society, and the Brewers' Exhibition at the Royal Agricultural Hall in London

In 1934, four local cidermakers were granted off-licences to sell their own cider under the National Mark Scheme, and the Linden Lea Cider Company was born, including the winner of the All-England National Mark Cider Award of 1932, Harry E. R. Warren,

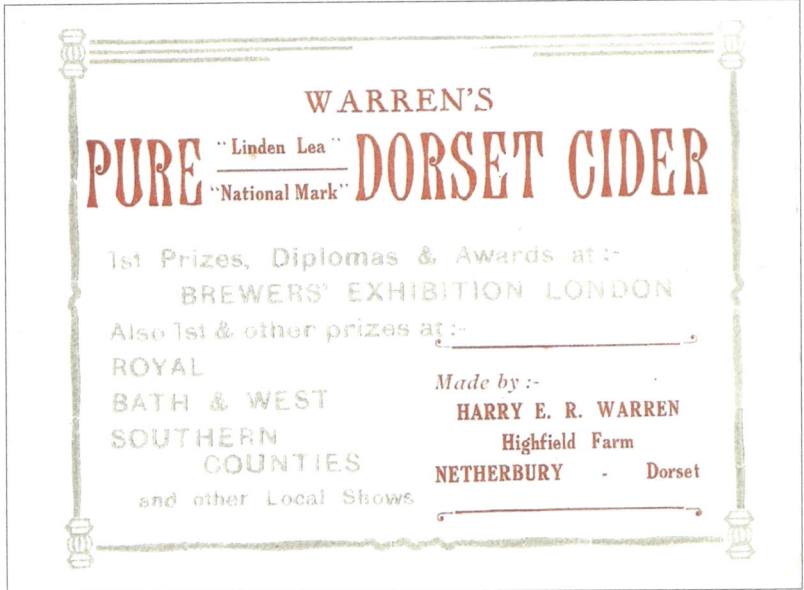

An early label from a bottle of Warren's cider in the 1930s.

with his neighbours William C. Oliver of Netherbury, Edward Smith of Stoke Abbott and James W. Best of Hincknowle, Melplash. The grant of their freedom to trade caused much opposition from local licensees, who saw them as worrying competition in the hard inter-war climate.

Four generations of the Best family have kept the orchards at Hincknowle since James Best bought the farm in 1922. (James was one of ten sons from a family that could trace its ancestry back to William Draper Best, the 1st Baron Wynford, of Wynford Eagle, a close personal friend of Captain Thomas Masterman Hardy of Trafalgar fame.) He was enthusiastically involved with Long Ashton and the Bath and West Society, where he judged both cider and sheep-shearing at the Shows.

James Best wrote about his orchards at Hincknowle over the years, logging the names, replacement trees and experimental introductions under the guidance of Long Ashton. He was keen to adopt recommended vintage cider apples, and began planting in

Harry Warren's price leaflet. Slightly sparkling 'Champagne' cider sold in heavy corked bottles under the National Mark Scheme.

earnest in 1928 by buying Kingston Black trees from Long Ashton for 7/6 each for his Muddicombe orchard, hidden away down the lane to Loscombe. He made the following detailed notes on flavours and blending of his varieties:

Balanced: Kingston Black and Cap of Liberty.
Sharp: Reinette Obry, Crimson King or Kings Favourite.
Sweet: Sweet Alford.
Bitter sweet: Dabinett, Bulmers Norman and Knotted Kernel.

To make a good cider, mix in equal proportions the apples of sharp, sweet and bitter sweet.
All eating apples and all cookers are sharp apples.

Encouraged by the development of his orchards and the success of his own cider in show competitions, James gave up cattle and sheep farming in favour of serious cidermaking with the Linden Lea Cider Company.

Harry Warren, champion cidermaker

Harry Warren, winner of many Long Ashton prizes, achieved a further pinnacle of success by winning the coveted McCreath Cider Cup of 1938/9. The prize was named after William McCreath, Robert Neville Grenville's right-hand man and expert cidermaker in the early days at Butleigh Court. It was awarded to the entrant earning the highest number of prizes in the year at the prestigious Bath and West Show, the Royal Show, the Brewers' Exhibition at Olympia, and the Imperial Fruit Show at Bristol. Harry Warren was up against strong names like Sheppy of Wellington, Pullin Bros of Compton, Greenfield of Bristol, Quantock Vale Cider from Bridgwater, Gloucestershire Cider at Wickwar, Dartington Hall in Devon and others. His McCreath Cider Cup achievement put Dorset cider on the map, along with the best of the West.

Harry Warren's orchards, in those days at Highfield Farm, Hatchlands and Perhay, held trees of Buttery Door, Best Bearer, Golden Ball, Mount Seedling and Woodbine. Close by in Stoke Abbott, his business partner Edward Smith kept a few different, complementary ones: Crimson King, Ironsides and one of his own, Smith's Sweet, together with several Somerset Crabs.

Harry Warren was an astute businessman, one of the first in the area to have mains electricity installed. He sent Hubert, the youngest of his three sons to Long Ashton to learn the art of cidermaking and blending. As much Linden Lea cider as possible was sold carbonated and in 'champagne' bottles, but the majority was sold in 9 gallon casks or as 4½ gallon pins. Harry continued to note Linden Lea cider sales and apple harvests in his diaries well into the 1950s. He once remarked, 'I take pride in my work because I consider it to be an ancient craft that must be preserved.'

The company traded highly successfully from 1924 until Harry's death in 1962. Hubert continued the business for years on a smaller scale, and it was in his orchard in 1998 that he proudly showed me his last real Dorset cider apple tree: a Golden Ball.

Hubert Warren testing the CO_2 pressure in a cider barrel.

Long Ashton's legacy

In 1939, with the start of the War, so much of Long Ashton's cider work was dropped. The celebrated cider competitions ended. Work was diverted to the Government Growmore Campaign, the land was turned to potatoes and cereals and advice focused on growing vegetables. Compulsory ploughing turned many farm orchards into arable or pasture, and no further fruit planting was allowed. Dorset saw the return of fields of flax and hemp. The cost of both wages and the mechanisation needed for increasing production put pressure on cidermaking. Energies had to be directed towards the most profitable options and many orchards saw their last cider harvest. Traditional orchards became unprofitable. Many good orchards on the outskirts of Dorset's villages became building sites. Fewer farmers made cider and many of those that did, made it for the often idiosyncratic taste of the ageing local male community. The final nails were put into the coffin by government grants to grub-up land for cereal crops.

But Long Ashton's work and instruction had achieved what was needed: ciders with 'more refined' flavour and a more profitable product for the industry. Farmers in all the cider counties had been persuaded to make new plantings. In the early years, they would have been Professor Berty Barker's tried and tested vintage apple varieties, like Foxwhelp, many of the well-known Somerset bittersweets, Kingston Black and even a few French introductions such as Bulmers Norman and Michelin. Dorset farmers were encouraged to plant more sharp varieties, including (thankfully) the award-winning Golden Ball.

Our cider industry was revitalised again in the 1970s. Cider was promoted once more as a national drink, often now by television advertising. Long Ashton began again with the essential work of selecting, trying and testing heavy-cropping cider varieties suitable for the burgeoning of bush orchards that were to supply the raw material to Bulmers, Showerings and the Taunton Cider Company. Most were Somerset bittersweet varieties like Yarlington Mill, Tremletts Bitter and Chisel Jersey, together with a large proportion of Michelin and Dabinett trees.

By far the biggest acreage of the commercially designed, closely planted new orchards were in Somerset and Herefordshire. Very few were established in Dorset, but those that were, are still significant contributors to cidermakers large and small. The Best family's orchards in and around Melplash, those planted just south of Beaminster by the Wood family, another big orchard in Waytown, a beautiful one near Cattistock, and another large group of orchards on the Dorchester road south of Sherborne, belonging to the Baxter family, are all still highly visible reminders of that golden age of commercial cider revival. Some of our more modern Long Ashton cider selections such as Three Counties, Hastings, Prince William, Lizzy and Gilly have joined them more recently.

Although Long Ashton's influence on the quality, popularity, marketability and reliability of Dorset cider was largely beneficial, the development has not been without a downside. With the burgeoning of fruit from a limited selection of varieties, ciders

A new generation of small-farm cidermakers at Monkton Wyld Cider.

coming from those bush plantings have a similar character and flavour, inevitably leading to the sad loss of much of the regional distinctiveness of local farm ciders. How lucky we were to find a few trees that survived unscathed in Dorset's deepest country districts. Having contributed in a leisurely way to make cider as the local customers wanted it, some good and some bad, long may these distinctive flavours live on to add something different to Dorset's own cider.

Very sadly, after 100 years of important horticultural science contribution, Long Ashton Research Centre was forced to close down and virtually all work and experimentation for cider orcharding ceased. I was the last cider pomologist employed there, and stayed till the doors closed in 2003. The whole site – buildings, and valuable land close to Bristol (belonging to the University) – was sold for development and some of the proceeds were put into trust for future work on apples at Bristol Centre for Agricultural Innovation.

As part of their ongoing research, geneticists based at the University of Bristol's School of Biological Sciences, led by Professor Keith Edwards (an ex-Long Ashton wheat geneticist) have been able to use the fund to develop a genotyping system, similar to that used for human DNA fingerprinting, which can rapidly and easily identify apple varieties. The University group, with help from John Thatcher of Thatcher's Cider in Somerset, and myself (now working independently), have sampled and analysed all the apple varieties of the National Fruit Collection at Brogdale in Kent, together with those from several Somerset orchards and various collections, private gardens and (most importantly) all our newly discovered Dorset apple trees. There is now an impressive database with several thousands of apple 'fingerprints', so useful for identifying apple trees, especially those whose names have been lost and forgotten in the past. Unfortunately, the techniques that they use are quite different and the results can't be directly compared. But it is early days in the world of genomics, with advances coming on apace.

The information that the two schemes have come up with so far is already telling us so much about the ancestry of our apple varieties: where they came from, who their parents were and which ones could be their offspring! The evolution of apple varieties through the ages is being slowly revealed and not unsurprisingly, some of the evidence backs up our hypothesis that the ancient Celtic legends of Dorset might indeed be true!

Glossary

bittersharp: a cider apple with juice that has more than 0.2% tannin and more than 0.45% malic acid.

bittersweet: a cider apple with juice that has more than 0.2% tannin and less than 0.45% malic acid.

blet, bletting: left to over-ripen, even to rot, before pressing.

budding: propagating a bush or tree by inserting a bud into a rootstock or young tree.

codlin: an ancient type of dual-purpose cooking apple.

costard: a cooking apple of medieval origin.

cultivar: a formal term for a variety of fruit or plant that will always be true to type.

diploid: plant variety containing the normal two sets of genes.

dual purpose: an apple suitable for eating, cooking and cidermaking.

fastigiate: a tree with a narrow upright growth habit.

genetic marker: a piece of DNA coding selected for identification.

genotype: the DNA coding specific to a single cultivar.

glass-to-glass: a blindfold tasting of cider to select the best impartially.

graft: propagate a bush or trees by inserting stem onto branch or trunk.

king fruit: the first fruit to set in a cluster is often bigger and a different shape from the others.

lenticel: one of many raised pores in the stem of a woody plant that allows gas exchange between the atmosphere and the internal tissues.

malic acid: organic compound that is made by all living organisms and contributes to the sour taste of fruits.

malolactic ferment: secondary fermentation that often occurs naturally, where malic acid develops into softer tasting lactic acid.

modern cider: low-tannin, easy-drinking cider, often on the sweet side.

Opposite: Cider apples, *c.*1989, by James Ravilious.

phenols: a class of chemical compounds consisting of one or more hydroxylgroups, synthesized industrially and produced by plants and microorganisms. Phenols are more acidic than typical alcohols.

pipe: large wine cask, perhaps 126 gallons or 2 hogsheads (478 litres).

pitchers: trees that will root from cuttings.

pleaching: the joining of stems or branches through grafting.

pomace: dry or pulpy residue of material, such as apples after pressing.

puncheon: large cask of varying capacity, usually 80 gallons (304 litres).

racking: drawing off dead yeast and other impurities from a fermenting liquid from.

ropiness and sickness: problems caused by bacterial contamination.

scab, mildew and canker: debilitating apple diseases.

scratter mill: for grinding apples ready for pressing.

sharp: acid tasting cider apple, usually with more than 0.45% malic acid in the juice.

specific gravity (SG): common way of expressing the sugar content of juice. The higher it is, the more alcohol is produced during fermentation.

stem builder: strong variety grown for two years before grafting with a cider variety.

sweet: cider apple that has plenty of sugar in its juice but little or no tannin or acidity.

tannin: term for one of the many complex bitter or astringent tasting molecules found in cider apples.

triploid: plant variety containing three sets of genes, one more than the normal two (diploid).

variety: informal term for a cultivar of a fruit or plant.

verjuice: unripe grape juice, usually for cooking purposes.

vernalise: allow seeds or buds a period of winter cold to release their dormancy.

vintage apples: those with the best quantity and quality of juice.

wilding: seedling apple tree of unknown parentage that is growing wild, especially by the wayside.

References

Boré, J. M. and Fleckinger, J. (1997) *Pommiers à Cidre*, INRA France.

Bryan, Guy (2004) *Wodetone – A Wooded Place: A History of Wootton Fitzpaine*, Creeds, Bridport.

Bunyard, Edward (1920) *Handbook of Hardy Fruits, Apples & Pears*, London.

Common Ground Book of Orchards (2000)

Copas, Liz (1999) *Somerset Pomona*, Dovecote Press.

Copas, Liz (2013) *Cider Apples, The New Pomona*, Somerset.

Creed, Sylvia (1987) *Dorset's Western Vale*, Sherborne.

Crowden, James (2008) *Ciderland*, Birlinn, Edinburgh.

Godwin, H. (1956) *History of the British Flora*, Cambridge University Press.

Higg, Robert (1886) *The Apple & Pear as Vintage Fruits*, Hereford.

Hinton, David (1998) *Saxons & Vikings*, Dovecote Press.

Hogg, Robert (1884) *The Fruit Manual*, London.

Juniper, B. and Mabberley, D. (2006) *The Story of the Apple*, Timber Press).

Kerr, Barbara (1968) *Bound to the Soil*, John Baker.

Knight, Thomas (1811) *Pomona Herefordiensis*, London.

Long Ashton Research Station, 100 years (2003), Bristol.

Mortimer, J. (1708) *The Whole Art of Husbandry*, London.

Pennington, Winifred (1969) *The History of British Vegetation*, Hodder.

Pollard, A. (1956) *Cider Fruit Production: Pamphlet 24*, Bath & West Society.

Roach, F. A. (1985) *Cultivated Fruits of Britain*, Blackwell.

Royal Horticultural Society (1935) *Apples and Pears, Report RHS Conference 1934*, RHS.

Scott, J. (1878) *Scott's Orchardist*, London.

Smith, Muriel (1971) *National Apple Register*, Ministry of Agriculture.

Index

Acknowledgements

Many people have supported our project along the way, especially James Crowden, Tom Munroe, Rupert Best and Nick's long suffering wife, Dawn. Also, Alan Symes, Graham Davies, Tim Beer, Alasdair Warren, Charlie Newman, Russel Crocker, Johnny Bowden, Jon Sibthorpe, Richard Meunier, the Worshipful Company of Fruiterers, Nichola Butler, Sylvia Creed-Castle, Nick Gray, Tim Connor, Chuck Wilmott and friends, Margaret Morgan Grenville, Elizabeth Bletsoe, Brian Earl, Richard Johnson, Keith Edwards, Bob Imlach, Bob Chaplin and Nick Mann. A special thanks to Adam's Apple, Dorset Nectar, Dorset Star, Isaacs, Purbeck Cider, Twisted Cider and West Milton Cider Co.. Thanks to all those who told us where to find the apple trees way back at the start, and to all those who supported the making of this book.

Nick and Liz, 2022

Opposite: Cattle grazing, Powerstock, Dorset, *c.*1989, by James Ravilious.

Oliver Rackham Library
THE ASH TREE
ANCIENT WOODS OF THE HELFORD RIVER
ANCIENT WOODS OF SOUTH-EAST WALES

Richard Mabey Library
NATURE CURE
THE UNOFFICIAL COUNTRYSIDE
BEECHCOMBINGS
GILBERT WHITE: A BIOGRAPHY

Nature Classics
WANDERERS IN THE NEW FOREST *Juliette de Baïracli Levy*
THROUGH THE WOODS *H. E. Bates*
MEN AND THE FIELDS *Adrian Bell*
THE ALLOTMENT *David Crouch and Colin Ward*
ISLAND YEARS, ISLAND FARM *Frank Fraser Darling*
AN ENGLISH FARMHOUSE *Geoffrey Grigson*
THE MAKING OF THE ENGLISH LANDSCAPE *W. G. Hoskins*
A SHEPHERD'S LIFE *W. H. Hudson*
WILD LIFE IN A SOUTHERN COUNTY *Richard Jefferies*
FOUR HEDGES *Clare Leighton*
DREAM ISLAND *R. M. Lockley*
RING OF BRIGHT WATER *Gavin Maxwell*
COPSFORD *Walter Murray*
THE FAT OF THE LAND *John Seymour*
IN PURSUIT OF SPRING *Edward Thomas*
THE NATURAL HISTORY OF SELBORNE *Gilbert White*

Field Notes & Monographs
AUROCHS AND AUKS *John Burnside*
ORISON FOR A CURLEW *Horatio Clare*
SOMETHING OF HIS ART: WALKING WITH J. S. BACH *Horatio Clare*
BROTHER. DO. YOU. LOVE. ME. *Manni Coe and Reuben Coe*
HERBACEOUS *Paul Evans*
THE SCREAMING SKY *Charles Foster*
THE TREE *John Fowles*
NEMESIS, MY FRIEND *Jay Griffiths*
TIME AND PLACE *Alexandra Harris*
ELOWEN *William Henry Searle*
EMPERORS, ADMIRALS AND CHIMNEY SWEEPERS *Peter Marren*
DIARY OF A YOUNG NATURALIST *Dara McAnulty*
LOVE, MADNESS, FISHING *Dexter Petley*
THE LONG FIELD *Pamela Petro*
SHALIMAR *Davina Quinlivan*
LIMESTONE COUNTRY *Fiona Sampson*
SNOW *Marcus Sedgwick*
WATER AND SKY, RIDGE AND FURROW *Neil Sentance*
BLACK APPLES OF GOWER *Iain Sinclair*
ON SILBURY HILL *Adam Thorpe*
GHOST TOWN: A LIVERPOOL SHADOWPLAY *Jeff Young*

Anthology & Biography
ARBOREAL: WOODLAND WORDS *Adrian Cooper*
MY HOUSE OF SKY: THE LIFE OF J. A. BAKER *Hetty Saunders*
CORNERSTONES: SUBTERRANEAN WRITING *Mark Smalley*
NO MATTER HOW MANY SKIES HAVE FALLEN *Ken Worpole*

Little Toller Books
w. littletoller.co.uk   e. books@littletoller.co.uk